essentials

essentials liefern aktuelles Wissen in konzentrierter Form. Die Essenz dessen, worauf es als „State-of-the-Art" in der gegenwärtigen Fachdiskussion oder in der Praxis ankommt. *essentials* informieren schnell, unkompliziert und verständlich.

- als Einführung in ein aktuelles Thema aus Ihrem Fachgebiet
- als Einstieg in ein für Sie noch unbekanntes Themenfeld
- als Einblick, um zum Thema mitreden zu können

Die Bücher in elektronischer und gedruckter Form bringen das Fachwissen von Springerautor*innen kompakt zur Darstellung. Sie sind besonders für die Nutzung als eBook auf Tablet-PCs, eBook-Readern und Smartphones geeignet. *essentials* sind Wissensbausteine aus den Wirtschafts-, Sozial- und Geisteswissenschaften, aus Technik und Naturwissenschaften sowie aus Medizin, Psychologie und Gesundheitsberufen. Von renommierten Autor*innen aller Springer-Verlagsmarken.

Yvonne Tafelmaier · Guido Bataille
Viola Schmid · Andreas Taller
Manuel Will

Methods for the Analysis of Stone Artefacts

An Overview

 Springer

Yvonne Tafelmaier
State Office for Cultural Heritage
Baden-Württemberg
Esslingen, Germany

Guido Bataille
State Office for Cultural Heritage
Baden-Württemberg
Blaubeuren, Germany

Viola Schmid
Austrian Archaeological Institute
Austrian Academy of Sciences
Vienna, Austria

Andreas Taller
Ur- und Frühgeschichte & Archäologie
des Mittelalters
Eberhard Karls Universität Tübingen
Tübingen, Germany

Manuel Will
Ur- und Frühgeschichte & Archäologie
des Mittelalters
Eberhard Karls Universität Tübingen
Tübingen, Germany

ISSN 2197-6708   ISSN 2197-6716   (electronic)
essentials
ISBN 978-3-658-39090-7   ISBN 978-3-658-39091-4   (eBook)
https://doi.org/10.1007/978-3-658-39091-4

What You Can Find in This *Essential*

- A compact overview of different methods for the analysis of stone artefacts
- Well-understandable explanations of the procedures of analysis
- Precise explanations of terms and definitions
- A critical examination of the advantages and disadvantages of the respective methodological approaches
- Further literature, if a deepen unterstanding of the topic or for partial aspects is desired

Contents

Introduction

During the Stone Age, knapped stone artefacts represent the most abundant category of all archaeological finds in most sites and provide the largest reservoir of potential knowledge for reconstructing this phase of human history. Stone artefacts, however, are not static objects. In prehistoric archaeology, until well into the twentieth century, only the end product, and thereby its final form and design, was used as a source of information. Stone artefacts played a particularly important role in determining the age of the associated find layer and thus of the archaeological assemblage. They served as so-called *fossils directeurs* (literally "*index fossils*"). This term is borrowed from geology and refers to artefact forms that only occur in certain epochs. Prehistorians have composed so-called type lists in order to distinguish lithic artefacts from one another on the basis of morphological differences (Bordes, 1950). From the 1980s onwards, especially in France increasing efforts were made to better comprehend the production process of lithic artefacts (Boëda, 1988, 1994, 1995; Boëda et al., 1990; Pelegrin, 1990, 1995). Technology (*conceptual knowledge*) on the one hand and technique (*practical implementation*) on the other hand played a key role. The interplay of both levels is the main focus of interest of technological investigations. Indeed, beyond their usefulness for dating, lithic artefacts offer a wide range of additional possibilities for gaining knowledge of the deep past (Fig. 1.1). For example, the analysis of specific production methods allows to draw conclusions about the technological concepts in artefact production known and practised in past communities; the study of raw materials, on the other hand, allows inferences on mobility and contacts of prehistoric human groups. A comprehensive study of lithic artefacts by various methodological approaches therefore holds enormous potential for reconstructing the cultural evolution of prehistoric hunter-gatherers and the everyday life of past societies.

Aspects of human behavior

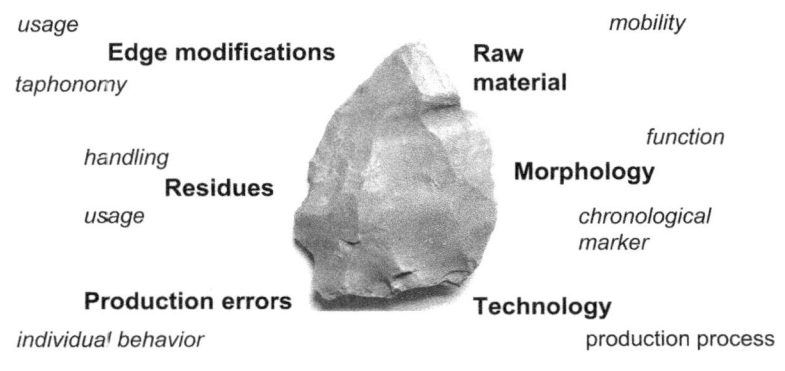

usage   mobility
Edge modifications   **Raw**
taphonomy   **material**

                                                          function
handling   **Morphology**
Residues
usage   chronological
                                                         marker

Production errors   **Technology**
individual behavior   production process

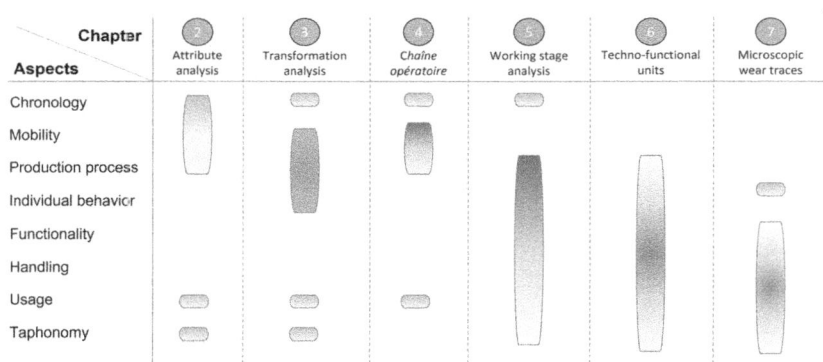

Fig. 1.1 Lithic artefacts are a rich source of information. Different characteristics allow drawing conclusions about a wide spectrum of potential research questions. These aspects include questions about the everyday life of Stone Age people, taphonomic processes, and general chronological and cultural classifications. The lower part of the figure shows which methods are particularly suitable for which questions and in which chapters they are introduced

In this short overview of different methods for analysing lithic artefacts used worldwide, we aim to make students and colleagues aware of these possibilities and to highlight their added value. Although the methods are presented separately for didactic reasons, the authors of this work share the common conviction arising from our teaching and research activities at German universities that the diverse approaches can and should be combined. The impetus and background for writing this book is based on the teaching concept implemented for many years for the study of stone tools at the University of Tübingen in the Department of Early Prehistory and Quaternary Ecology. Here, various lecturers teach students on the different methods in lectures and practicals in addition to a common basic course of "Artefact Morphology". In the individual chapters, the different analytical methods are briefly described, historically sketched, examples of application explained, and strengths and weaknesses addressed. According to our respective courses in teaching, we have divided the chapters as follows:

Chapter 2: Attribute Analysis (Manuel Will).
Chapter 3: Transformation Analysis (Guido Bataille).
Chapter 4: *Chaîne opératoire* Approach (Viola C. Schmid).
Chapter 5: Working Stage Analysis (Yvonne Tafelmaier).
Chapter 6: Analysis of techno-functional units (Yvonne Tafelmaier).
Chapter 7: Microscopic Use-Wear Analysis (Andreas Taller).

Of course, not all methods can be considered within the scope of this brief overview. We present those which, in our view, are the most important and practicable. Techniques for the documentation of archaeological objects, i.e. the creation of drawings (cf. Hahn, 1992), photographs as well as three-dimensional models are not part of this work; neither are in-depth statistical analyses.

Basic knowledge of stone artefacts is required for reading this book. Numerous technical terms cannot be explained in detail here. For a first introduction, German-speaking people should refer to the book "Erkennen und Bestimmen von Stein- und Knochenartefakten – Einführung in die Artefaktmorphologie" written by Joachim Hahn in 1991. This standard work has not lost its validity. The anthology "Steinartefakte" (Stone artefacts) edited by Harald Floss in 2012, which deals mainly, but not exclusively, with the classification of stone artefacts, is also recommended reading. A highly appreciated international standard work is the English edition "Technology and terminology of knapped stone" by Inizan et al. (1999), which includes a useful multilingual vocabulary. Further pertinent introductory textbooks in English providing basic (and advanced knowledge) include "Lithics - Macroscopic Approaches to Analysis" (Andrefsky, 2005), "Lithic Analysis" (Odell, 2004) and "A Record in Stone" (Holdaway & Stern, 2004).

Attribute Analysis

2

2.1 Introduction

On the most fundamental level, analyses of stone artefacts can be divided into reductionist and holistic approaches. The **holistic**, or **hermeneutic**, approach considers a stone artefact (or an assemblage) in its entirety as an analytical object. This holistic approach focuses on the traits of a lithic artefact/assemblage in their interaction and totality, including typological and technological classifications of lithic objects (see Chap. 4). The **reductionist**, or **atomising**, approach, on the other hand, apportions an assemblage into individual lithic artefacts, and lithic artefacts into individual elements (attributes or traits) with the latter constituting the basic analytical units. An attribute analysis thus breaks down the complex morphology and metrics of stone artefacts into their component parts in order to define, measure, address and evaluate them individually. The aim and epistemological interest of the method is to reconstruct the methods and techniques of stone artefact production occurring within an assemblage or sample of lithic artefacts based on the quantitative analysis of these attributes, to generate verifiable interpretations on reductions sequences and techno-economic behaviour, and to test specific hypotheses.

In order to draw meaningful and directed interpretations about the past on the basis of distinguishable traits on stone artefacts, attribute analysis relies on theoretical and empirical reference points. This knowledge on specific knapping features derives from the general physical principles of **fracture mechanics**, which investigates the laws underlying stone flaking and thus allows sound inferences (Cotterell & Kamminga, 1987). **Experiments**, both controlled and replicative studies, provide further knowledge on the relationship between specific analytical

© The Author(s), under exclusive license to Springer Fachmedien Wiesbaden GmbH, part of Springer Nature 2022
Y. Tafelmaier et al., *Methods for the Analysis of Stone Artefacts*, essentials, https://doi.org/10.1007/978-3-658-39091-4_2

attributes and the methods and techniques of stone knapping that are to be reconstructed (Whittaker, 1994; Pelcin, 1997; Pelegrin, 2000; Dibble & Rezek, 2009).

Attribute analysis has a played dual role in lithic analysis: firstly, as a **methodological school** with a quantitative-statistical focus and, secondly, as a general **analytical approach** and pure **catalogue of traits**. As a methodological school, *attribute analysis* is a dominantly Anglo-Saxon phenomenon, with important representatives primarily in the USA and Australia, which has permeated many other regions and research traditions. This school is characterized by a focus on measurable, standardized **quantification** of individual attributes and their statistical analysis. Subjective, interpretive and evaluative elements of other methods are largely avoided in favour of a more objective, replicable and transparent recording of stone artefacts. Attribute analysis, however, is intentionally placed as the first chapter in this book, since in principle all subsequent methods, as well as classifications and typologies of stone artefacts, are based on the (implicit or explicit) use of observable traits on lithic artifacts.

2.2   History of Research

The use of traits to classify stone artefacts already characterises the origins of Stone Age research. The famous typologies of stone tools developed in the twentieth century (Bordes, 1961) are also based on the mostly unspecified combination of attributes, for example the form and extent of retouching. An explicit focus on individual attributes on lithic objects and their quantitative recording can be grasped especially in Anglo-Saxon research communities since the 1960s and 1970s. Roots for these developments can be found in the emerging New Archaeology with its desideratum for the use of more objective and statistical methods, a focus on variability and the analysis of all products of the manufacture chain, as well as a closer link with the experimental archaeology of stone knapping and important work on fracture mechanics (Speth, 1972; Dibble & Whittaker, 1981; Cotterell et al., 1985; Cotterell & Kamminga, 1987). Although rarely clearly formulated as a methodological school, influential publications (Fish, 1981; Sullivan & Rozen, 1985; Shott, 1994; Tostevin, 2003, 2012; Dibble & Rezek, 2009), PhD theses (e.g. Magne, 1985), and textbooks on stone artifact analysis with a strong or exclusive focus on quantitative analysis of traits (Odell, 2004; Holdaway & Stern, 2004; Andrefsky, 2005; Shea, 2013) point to a prevalence of attribute analysis as the dominant direction in the Anglo-Saxon research community. This is not the case in other countries and contexts. As one example, in the German-speaking world, attribute analysis plays a minor role as a methodological school. However, a clear reference to the recording of individual attributes and their quantitative analysis based on individual stone artefacts can be detected in a number of important methodological publications (Kerkhof & Müller-Beck, 1969; Auffermann et al., 1990; Drafehn et al., 2008), textbooks (Hahn, 1991) and monographs on the study of key assemblages from the German Palaeolithic and Neolithic (e.g. Hahn, 1988; Zimmermann, 1988). Similar as in other contexts, attribute analysis is here mainly seen as one instrument among several in the toolbox of methods and combined with them.

2.3   Approach (Method)

In principle, attribute analysis does not examine asemblages in their totality. Instead, individual stone artefacts form the main analytical units, and these individual artefacts are examined via selected metric and morphological attributes. Only after data recording on individual pieces, the assemblage is analyzed via quantitative and statistical methods using either the entirety or selected samples of the collected data. There is no fixed scheme or cookbook recipe for attribute analysis. The reason for this lies in the varying thematic and temporal context of potential studies, the nature of the assemblage, and the underlying question and scope of the analysis. In any case, attribute analysis comprises the following work steps: (a) preparation and processing; (b) collection/recording of data; (c) analysis of data; (d) interpretation and contextualisation of results.

2.3.1 Preparation and Processing

Before starting an attribute analysis, the theoretical, thematic and concrete context of the assemblage as well as the corresponding objective, research questions, hypotheses and scope of the analysis need to be determined or considered. These factors significantly influence the following steps (b)–(d) mentioned above. Preparation includes an initial review or basic accounting of the assemblage, which includes a numerical overview of the stone artefacts, the diversity of the range of forms and raw materials, and any necessary processing work such as cleaning and individual labelling of stone artifacts. Subsequently, the researcher needs to decide which pieces are to be included in the attribute analysis, e.g. all existing artefacts or only pieces above a certain size ("size cut-off").

Only after this preliminary work can a **list of traits** be compiled. This list is based on the general framework mentioned above. It should record the morphology and metrics of the artefacts, enable interpretations about stone knapping and allow answering the posed research questions. On a basic level, different **attributes** are recorded for blanks, cores and tools (Figs. 2.1, 2.2, and 2.3). While there exists no single list of attributes applicable to all contexts, minimal lists have been produced by some authors (e.g. Shott, 1994). These attributes include weight, raw material, cortex percentage, bulb, number of dorsal negatives, type of platform, and basic measurements such as maximum dimension, length, width, and thickness. Other characteristics include basic features to record metrics and knapping techniques (Figs. 2.1 and 2.2). Lists and depictions of possible metric and discrete attributes can be found in the relevant literature and textbooks and are omitted here due to a lack of space (Auffermann et al., 1990; Hahn, 1991; Odell, 2004; Holdaway & Stern, 2004; Andrefsky, 2005; Shea, 2013).

Three steps are crucial in the creation of the specific recording scheme. (1) **Definition of** the attribute to be recorded. Clear, transparent and comprehensible definitions ensure replicability. (2) Establishing the **expression or state** of the attributes, including determination of the measurement level (discrete vs. continuous). A caetgory for not available (NA) should be included. (3) Reflection on the **utility and relevance** of the attribute to be included according to the context and goal of the analysis (i.e. the information content of the attributes). This helps in choosing attributes to be included, narrowing down the near endless range of recordable traits. The attribute list is reworked into a **recording grid** (Fig. 2.4) to allow simple collection of data per attribute. On this basis, a **database** with an in-

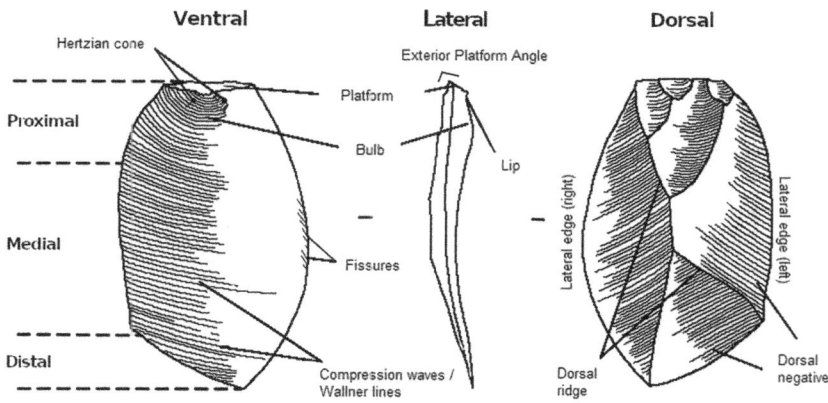

Fig. 2.1 Overview of terms and general discrete knapping traits on blanks. (Graphic: Manuel Will & Melanie-Larisa Peter, created on the basis of Inizan et al., 1999: Fig. 5)

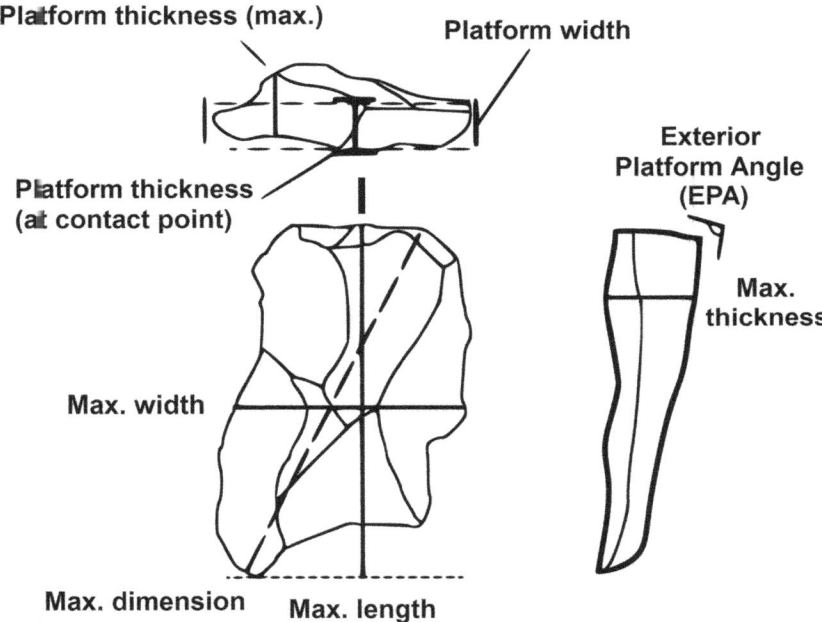

Platform thickness (max.)

Platform width

Platform thickness (at contact point)

Exterior Platform Angle (EPA)

Max. thickness

Max. width

Max. dimension **Max. length**

Fig. 2.2 Illustration of the most important metric attributes (measurements) on blanks. EPA = Exterior Platform Angle. (Graphic: Manuel Will & Melanie-Larisa Peter)

put mask is created. Although other programs (e.g. Excel) also allow straightforwarding recording, they lack key features of a relational database system (e.g. control of large amounts of data, consistency checks, efficient retrieval of data).

2.3.2 Collection/Recording of Data

The collection or recording of data always takes place at the level of an **individual stone artefact** and following the **selected attributes and attribute states** by entering them into the database. Individual pieces should therefore be clearly labelled and stored separately in individual finds bags. One after the other, the individual artefacts are first apprehended in general and examined for special features and possible problems for data collection (e.g. crusts on the surfaces). In the following, individual attributes are analysed one after the other and the researcher records the respective attribute state or takes the measurement. Subsequently, the inputs are

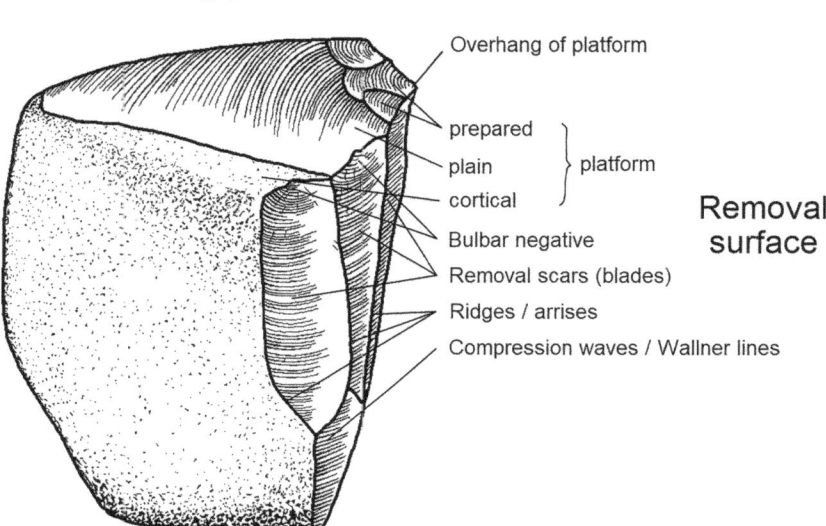

Fig. 2.3 Overview of terms and general discrete knapping traits on cores. (Graphic: Manuel Will & Melanie-Larisa Peter, created on the basis of Inizan et al., 1999: Fig. 20)

checked for correctness. Finally, the stone artefact is put back and marked as entered with an appropriate reference (e.g. a mark on the associated find bag), and the next piece is analyzed. The following guidelines should be followed in this process: The chosen approach and attribute list is consistently followed. Recording and measurements must be made accurately so that the data are reliable. Constant documentation through analog or digital notes, photography and drawing is helpful. The necessary instruments – magnifying glasses, lamps, scales, calipers, goniometers – should be checked for correct function and used in a consistent manner throughout data collection.

2.3.3   Analysis of Data

Data analysis is always carried out at the level of the **entire assemblage** or **selected samples**, never on the individual piece. Results of the data analysis are representative, quantitative summaries of the stone artefact assemblage according to indi-

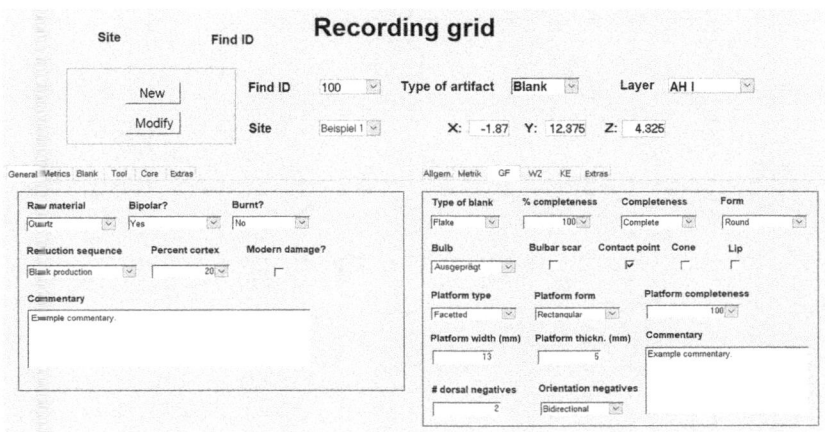

Fig. 2.4 Example for a recording grid for attributes on stone artefacts (here site Hoedjiespunt 1, South Africa). Attributes as headings (bold), attribute states are either selectable in a drop-down menu or to be entered freely in boxes. (Graphic: Manuel Will)

vidual attributes. Quantitative description of the assemblage of e.g. an archaeological layer consist of simple tabulations of numbers (n) and proportions (%) of categorical attributes (e.g., tool types, raw materials, basic shape types), descriptive graphs, or statistical parameters of continuous traits (e.g., average, maximum, and minimum length of blades or cores). The analysis of selected samples, such as examining all retouched artifacts or only blades, helps to answer specific questions. Data analysis in a relational database (Fig. 2.5) allows for the selection of any (combination of) attributes and samples via so-called queries.

Before each data analysis, all data need to be checked for correctness. Ensuing quantitative analysis follows the rules of descriptive and analytical statistics. Basically, quantitative data can be presented in **tables, graphs** and **statistical parameters** (Fig. 2.6). This transformation of the attribute data from the database into representational figures is done with the help of suitable calculation software such as Microsoft Excel or statistical programs such as SPSS or R. Additional procedures from analytical statistics allow testing of specific hypotheses on significant differences and associations (e.g. associations of raw materials with blanks; size differences between tool types). Proper handling of quantitative data is critical and relevant guidelines can be found in statistics textbooks.

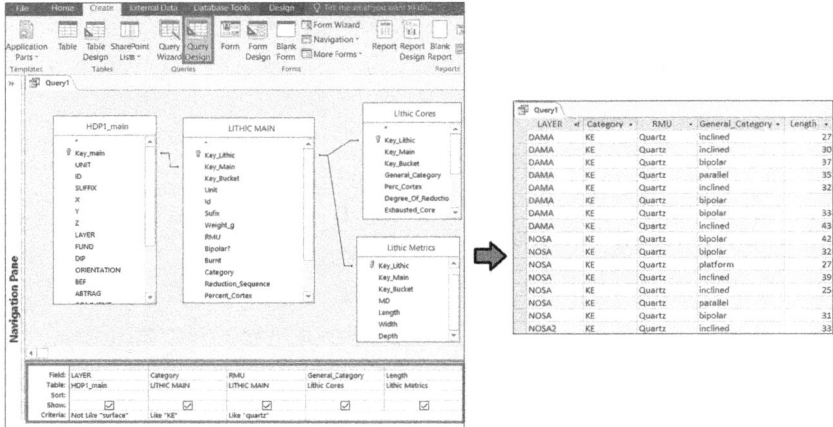

Fig. 2.5 Example data query for attributes on stone artefacts from a Microsoft Access database (here site Hoedjiespunt 1, South Africa). Left: Data query with linked relational tables and criteria of the query (bottom left), Right: Result display of the query. (Graphic: Manuel Will)

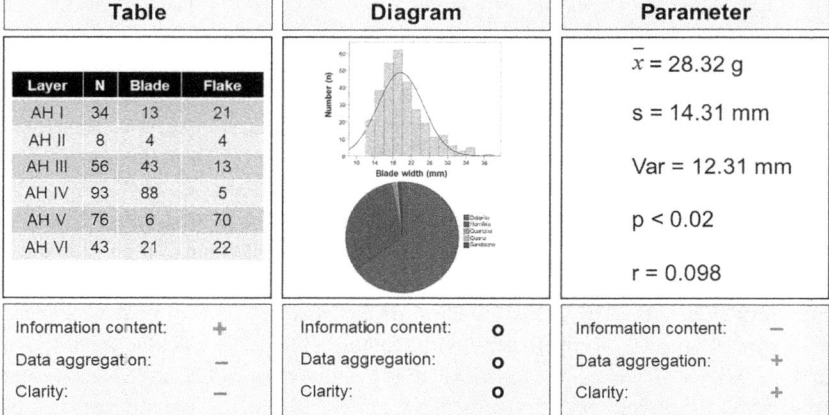

Fig. 2.6 Overview of the advantages and disadvantages to the form of presentation of quantitative data in tables, charts and statistical parameters with examples. (Graphic: Manuel Will)

2.3.4   Interpretation and Contextualisation of Results

The final step of attribute analysis lies in the interpretation of the quantitative results. Interpretations are principally based on the reference to the laws of fracture mechanics, experimental research on stone artefacts, and other published artefact analyses (e.g. comparison with results of other methods or studies). This linkage leads to robust interpretations regarding the reconstruction of knapping techniques and methods as well as reduction sequences and techno-economic behaviour of the specific assemblage or sample under study. Influencing factors such as the nature and context of the assemblage (e.g. surface collection; size-sorted assemblage) must be taken into account and critically examined. At the end of the analysis, postulated interpretations are further embedded with regard to the concrete research questions and the general topic of research.

2.4   Strengths and Weaknesses, When to Use, When Not to Use?

The strengths of attribute analysis lie in its high degree of **transparency, intersubjectivity** and **replicability**, as well as the comparatively low prerequisites (i.e. expert knowledge) for the analyst. In principle, attribute analysis can therefore be applied to any stone artefact assemblage, regardless of its temporal, spatial or cultural context. The weaknesses of the methodology concern the **sample size** required to enable meaningful quantitative analysis, and a more superficial description of artefacts that lacks the dynamism and detail of interpretative, holistic approaches. Attribute analysis is useful when a large assemblage is available (n > 100), a general overview is to be provided, and the research question is focused on comparative studies. Other or complementary approaches are more appropriate for small assemblages (n < 30) and in-depth studies of individual find categories (e.g. handaxes) or reduction methods that are described in the following chapters of this book. Ideally, these approaches should also be based on and/or combined with clear attribute lists.

Transformation Analysis 3

3.1 Introduction

Stone tools are the result of complex and varied manufacturing processes that are often fragmented in time and space. The initial knapping of a **raw piece**, e.g. the decortication of a raw nodule, may already have been carried out at the original **deposit**. Determining the method of core reduction concepts and, associated with this, the process of **core formatting as** well as the subsequent extraction of **blanks** (flakes, blades and bladelets) then took place, for example, at a **hunting station** that was occupied for a short period of time. At the same time, blanks prepared into different tool forms may have been carried along for activities in the future and deposited at different places. With the help of the **transformation analysis** (TA in the following), such temporally and spatially fragmented activities can be reconstructed and technological concepts adequately described. For this purpose, **artefact assemblages** of archaeological horizons are sorted into **raw material units** (RMU) on the basis of macroscopic, i.e. visible to the naked eye, characteristics of their **raw material** properties. The aim is to resolve the stone artefacts of an assemblage as far as possible into nodule-like units. Such lithic units, derived from an original raw piece, are referred to as **work pieces**. This research approach, which divides assemblages into lithic units based on the analysis of raw material properties, has methodological overlaps with the creation of **refitting units,** as the latter requires such presorting (Cziesla, 1986). The workpieces recognized via the sorting process can provide information about applied **technological concepts** of stone artifact production in individual assemblages. If necessary, the range of different techno-functional production strategies that are evident in different workpieces can be grouped into specific operational schemes. Thus, there are also over-

© The Author(s), under exclusive license to Springer Fachmedien Wiesbaden GmbH, part of Springer Nature 2022
Y. Tafelmaier et al., *Methods for the Analysis of Stone Artefacts*, essentials, https://doi.org/10.1007/978-3-658-39091-4_3

laps in terms of content and form with the approach of reconstructing **operational chains** (French: *chaînes opératoires*) taken from French social anthropology (cf. Chap. 4). In contrast to the **work step analysis** (cf. Chap. 5) and its French counterpart the *'méthode diacritique'*, which examines operational sequences by analysing individual artefacts, The TA reconstructs lithic reduction strategies on the basis of entire stone assemblages (Chabai et al., 2005, 2006). Thus, reduction sequences, i.e. specific action sequences and activities, can be reconstructed from the individual raw material units based on tradition and individual solution approaches.

3.2   History of Research

The TA was developed as a methodological concept by the Erlangen prehistorian W. Weißmüller (1950–2005) as part of his habilitation at the University of Erlangen on the basis of Middle Palaeolithic stone assemblages from the Sesselfelsgrotte (Bavaria) (Weißmüller, 1995). Weißmüller oriented himself on existing research approaches, such as the reconstruction of concepts of core reduction (e.g. Geneste, 1985; Boëda et al., 1990). The methodological starting point is the systematic analysis of raw material properties of in-situ lithic assemblages. In the German-speaking research community, G. Bosinski had already investigated the raw material properties of Middle Palaeolithic lithic assemblages in the 1960s and formed refitting and thus also import units on their basis (Bosinski et al., 1966). Based on G. Bosinski's work, H. Thieme made investigations on site-specific actions and defined activity zones (Thieme, 1983, p. 93 ff.). H. Löhr (1979), K. H. Rieder (1981/82), J. Hahn (1988), W. Roebroeks (1988), and N. J. Conard et al. (1998) also used macroscopic features of lithic artifact inventories to create workpieces and refitting units. In this way, the resolution of reconstructing human activities and sequences of actions that contributed to the accumulation of artifacts at a site increased. Weißmüller also devised a uniform and standardized methodological concept for the determination and evaluation of raw material units. This has been applied since the late 1990s, especially by German-speaking researchers, to investigate functional relationships between Palaeolithic as well as Mesolithic stone artefact assemblages and environmental parameters (Richter, 1997; Kind, 2003; Chabai et al., 2005, 2006; Böhner, 2008). In particular, as part of this and more recent work, the method was linked to an attribute analysis system that recorded techno-typological features of stone artefacts as well as raw material-specific attributes of raw material units (Uthmeier, 2004a, b; Bataille, 2017).

While the search for refitting sequences now has a decades-long tradition in international research and is part of the present standard repertoire for the analysis of lithic and organic artefacts, the methodology of workpiece sorting is only gradually beginning to gain acceptance as a standardised method on an international scale (e.g. Machado et al., 2016; Romagnoli et al., 2016; Romagnoli & Vaquero, 2015).

3.3   Transformation Analysis: A Method for Reconstructing Past Activities

The knapping of stone artefacts takes place in the context of a continuous sequence of actions and events, whereby material is split off in the course of individual steps and changed or transformed in form and function (Weißmüller, 1995, p. 13). The goal of the TA is to return the individual artefacts to an original raw material unit and to reconstruct the original reduction sequences. The central element of the TA developed by W. Weißmüller is therefore the formation of workpieces, i.e. the sorting of individual artefacts back to an original raw piece on the basis of common features of the raw material characteristics (Uthmeier, 2004a). In this way, the individual stages of the manufacturing processes of stone artefacts can be recorded in their technological, spatial and functional context.

The TA can be divided into two steps. (1) The first step is the sorting of stone artefacts according to macroscopic features into common raw material groups, which is followed by (2) the reconstruction of the physical transformation, at the site. In this way, it is possible to infer stages of reduction that took place prior to the occupation of the site as well as the subsequent export of lithic material (e.g. Uthmeier, 2004b; Bataille, 2010). Thus, it is finally possible to assign workpieces to different transformation classes in order to reconstruct specific transformation and reduction sequences. Technologically significant artifact types (e.g., primary crested blades), cortex coverage, and metric measures (cf. Chap. 2) thus provide information about the presence and absence of technological steps (e.g., the primary preparation of the core crest) that belong to a specific reduction sequence. Finally, building on these stages of TA, further interpretations can be made about applied technological concepts (Weißmüller, 1995), find and site contexts (Bataille, 2010), as well as site function and genesis (e.g., Uthmeier, 2004b; Kretschmer, 2005; Bataille, 2012).

In this way, analysts can reconstruct transformation processes carried out at the site, which are based on individual decisions and technological traditions (= **concepts**). By identifying missing objects, such as decortication blanks, crested blades or certain core types, which must have contributed to the formation of a raw material unit, statements can be made about previous events and, if possible, about activities directed towards the future (Uthmeier, 2004b). In a very concrete way, this approach attempts to "reconstruct the action that led to the formation of the find accumulation" (Weißmüller, 1995, p. 27). In this context, Weißmüller (1995, p. 28 ff.) refers to **conceptual reservoirs** derived from socially transmitted and individual experience from which stone knappers can draw.

3.3.1 Raw Material Analysis and Workpiece Formation

In the first step, stone artefacts are sorted back to original rock pieces according to their raw material properties (= raw material units). "Workpieces consisting of two or more artefacts of the same type" can be traced back to original lithic units (Weißmüller, 1995, p. 58). This workpiece formation is intended to reconstruct "the condition of the lithics on arrival at the site (= import condition)" as well as "the stages of transformation in the area of the site (= transformation stages)" (Fig. 3.1) (Weißmüller, 1995, p. 247). Artefacts that cannot be assigned to any other piece due to specific characteristics are classified as single pieces (Weißmüller, 1995, p. 58). Often such exotic mostly numerically small raw material units or single pieces originate from more remote raw material deposits (Bataille, 2010).

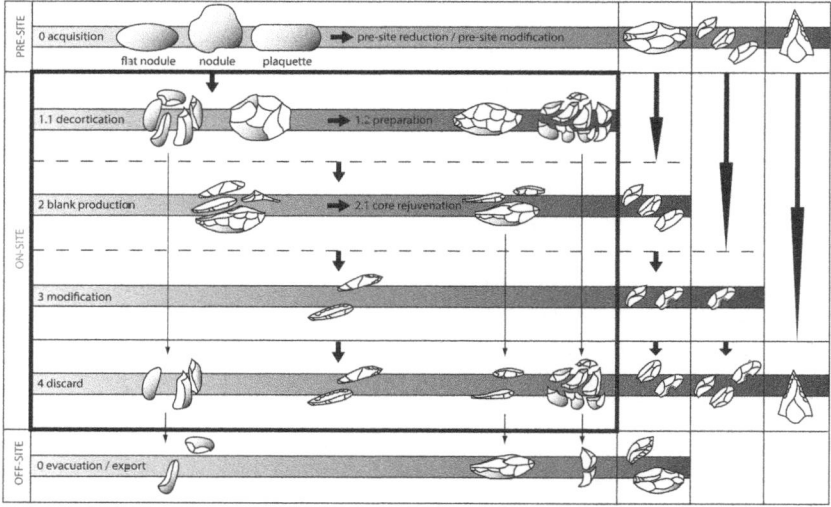

Fig. 3.1 Possible reduction chains and transformation sections of stone artifact production exemplified on Levallois reduction sequences as an example. (Graphic: Guido Bataille)

Flint is particularly suitable for raw material sorting, as this rock often exhibits pronounced structural and morphological variability. Basically, the more heterogeneous the examined raw material, the more suitable it is for high-resolution sorting on a workpiece basis. On the other hand, a high degree of patination as well as a large number of burnt pieces within an assemblage prevent a clear differentiation according to raw material characteristics. Such artefacts "which are not further addressable due to the raw material, the preservation or the too small size", must therefore be separated out as sorting residue (Weißmüller, 1995, p. 58). In the course of workpiece formation, the following object classes are generated on the basis of the sortability of a given lithic assemblage: single pieces, workpieces, varieties of a raw material outcrop and lithics belonging to a geological formation (Fig. 3.2).

The basis of raw material sorting are macroscopic characteristics of cortex and fracture plains (Fig. 3.2). These are colour and structure of cortex and raw material matrix, but also characteristic fossil inclusions, banding or cracks. Lists of raw material characteristics adapted to the respective raw material spectrum enable a systematic recording of raw material-specific variants (Uthmeier, 2004a) (Figs. 3.3

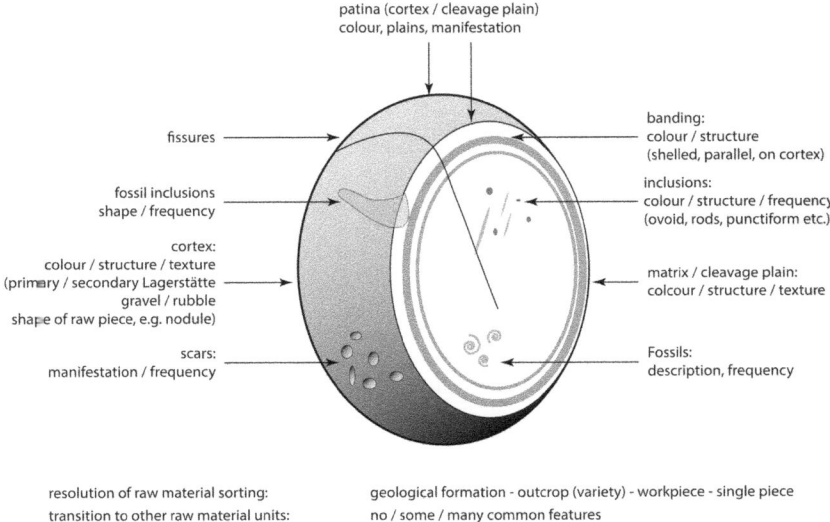

Fig. 3.2 The determination of different attributes and their expressions of lithic raw material is the prerequisite for a successful raw material sorting and transformation analysis. (Graphic: Guido Bataille based on Th. Uthmeier (2004a: Fig. 11–2))

SHEET 1 - RAW MATERIAL SORTING - MACROSCOPIC FEATURES

site: (name)   **GH:** (Geological Horizon)   **AH:** (Archaeological Horizon)   **RM:** (Raw Material)

number of artefacts: (per raw material unit)   **total weight:** (per raw material unit)

1. State of import of the raw material unit   3. remarks

1. resolution   2. similarities (to other raw material units)

single piece ☐   never   ☐
workpiece   ☐   occasional   ☐
variety   ☐   frequent   ☐

2. Cortex - attributes   without ☐   5. remarks

1. cortical remains   2. state   3. Lagerstätte

0%	☐	fresh	☐	primary	☐
< 25%	☐	scratchable	☐	residual	☐
< 50%	☐	still scratchable	☐	gravel	☐
< 75%	☐	smooth	☐	rubble	☐
< 100%	☐	scarred	☐	others	☐

4. colour

3. Raw volume   nodule ☐   plaquette ☐   not recognizable ☐
                              flat nodule ☐   others ☐

4. Cleavage plains - attributes   7. remarks
1. structure   2. glace   3. patina

crystalline	☐	glistening	☐	lacking	☐
abrasive	☐	shiny	☐	complete	☐
smooth	☐	matt	☐	partial	☐
others	☐	others	☐	upper / dorsal	☐
				lower / ventral	☐

4. banding   5. schlieren   6. colour

lacking	☐	lacking	☐
on cortex	☐	diffuse	☐
shelled	☐	contoured	☐
others	☐	others	☐

5. Macroscopic features - enclosures / fossils   5. structure   7. remarks
1. druses   3. enclosures   4. shape

present	☐	few	☐	punctiform	☐	sharp	☐
lacking	☐	many	☐	circular	☐	diffuse	☐
		lacking	☐	ovoid	☐	quarzitic	☐
2. fissures		others	☐	amorph	☐	others	☐
present	☐			rods	☐	6. colour	
lacking	☐			others	☐		

Fig. 3.3 Example of a recording sheet for raw material determination of a raw material unit. (Graphic: Guido Bataille)

SHEET 2 - TRANSFORMATION				
Site: (name)	**GH:** (Geological Horizon)	**AH:** (Archaeological Horizon)	**RMU:** (Raw Material Unit)	

1. Raw material unit

1. resolution		2. total artefacts
single piece	☐	
workpiece	☐	3. total weight
variety	☐	

4. Remarks

2. Transformation section

0. reconstructed	raw piece ☐	blank ☐	chunk ☐
import state	core ☐	tool ☐	others ☐

cortical state	cortical (100%)	partially cortical (<100 %)	non-cortical (0%)
1. Raw piece / core			
raw piece (nodule etc.)			
core			
2. Core preparation			
decortication blanks			
preparartion blanks			
3. Blank production			
blanks / waste (< 1-3cm)			
flake			
blade(max. width >/=12mm)			
bladelet(max. width < 12 mm)			
microblade (max. w. < 6 mm)			
lamellar burin spall / waste			
preparatory blank (crested)			
blank fragment			
chunk			
others			
4. Core correction / rejuvenation			
primary			
secondary			
5. Tool production			
modified piece (tool)			
modification waste			
tool fragment			
tool correction / rejuvenation			

3. Transformation class (acc. to Weißmüller 1995)

single piece		preparation/ correction		static objects of tool modification		blank obtainment/ tool modification	
Nw		Np		Ei		Nb	
Cw				TT		Cb	
Bw		Cc		Mi		Nm	
Tw				TM		Cm	

Fig. 3.4 Example of a recording sheet for determining the transformation of a raw material unit at the site. (Graphic: Guido Bataille)

and 3.4). The use of standardized color scales (e.g. Munsell Color Charts) is useful for determining colors and color nuances. Microscopic methods can be used to support the macroscopic analysis. Such analyses are particularly advantageous for the more precise determination of fossil inclusions etc. and in the context of checking the sorting quality.

3.3.2   Reconstruction of the Raw Material Transformation

The actual TA, i.e. the investigation of lithic transformation at the site, can be divided into two further steps: (1) the determination of import state and observable transformation stages per raw material unit and (2) the determination of the transformation section or transformation stage of the overall inventory based on these raw material units (Figs. 3.3 and 3.4). Another important point is represented by the "evacuation" (Weißmüller, 1995, p. 67 ff.) of lithic material, i.e. the absence of certain artifact categories (e.g. cores, crested blades, or waste from tool manufacture) in raw material units and assemblages (Fig. 3.1). Such absence can be attributed to post-depositional relocation, incompletely excavated find horizons, but also to deliberate export of certain artifacts (Chabai et al., 2006).

First, the recognized raw material units are categorized according to the presence of technologically meaningful artifact categories, the determination of the respective proportion of cortex remains as well as metric measures. In this context, Weißmüller defined 14 transformation classes, which allow statements on the state of import and the respective transformation section detectable at the site (Weißmüller, 1995, pp. 61 f., 67 ff.) (Fig. 3.5). This procedure serves the better understanding of the above mentioned spatially and temporally fragmented production chains of stone artefacts, in order to be able to reconstruct past activities. Weißmüller oriented himself on models of the chaîne opératoire approach of Middle Palaeolithic assemblages (e.g. Geneste 1988; Boëda et al., 1990).

The defined transformation classes are each designated by two Latin letters, the first providing information about the import state and the second about the reconstructed transformation at the site (Weißmüller, 1995: Fig. 21) (Fig. 3.5). For example, a distinction is made between workpieces that were brought in as a tool (T), blank (B), core (C), or raw nodule (N). The presence of specific "dynamic" categories (Weißmüller, 1995, p. 68), such as basic forms of core preparation and core correction, then provide information about the on-site transformation of imported raw material units. Based on the extent of cortex coverage of a raw material unit, it can be distinguished whether blanks (b) originate from imported raw nodules (Nb) or from cores (Cb). Such dynamic objects, i.e. those that can potentially be further

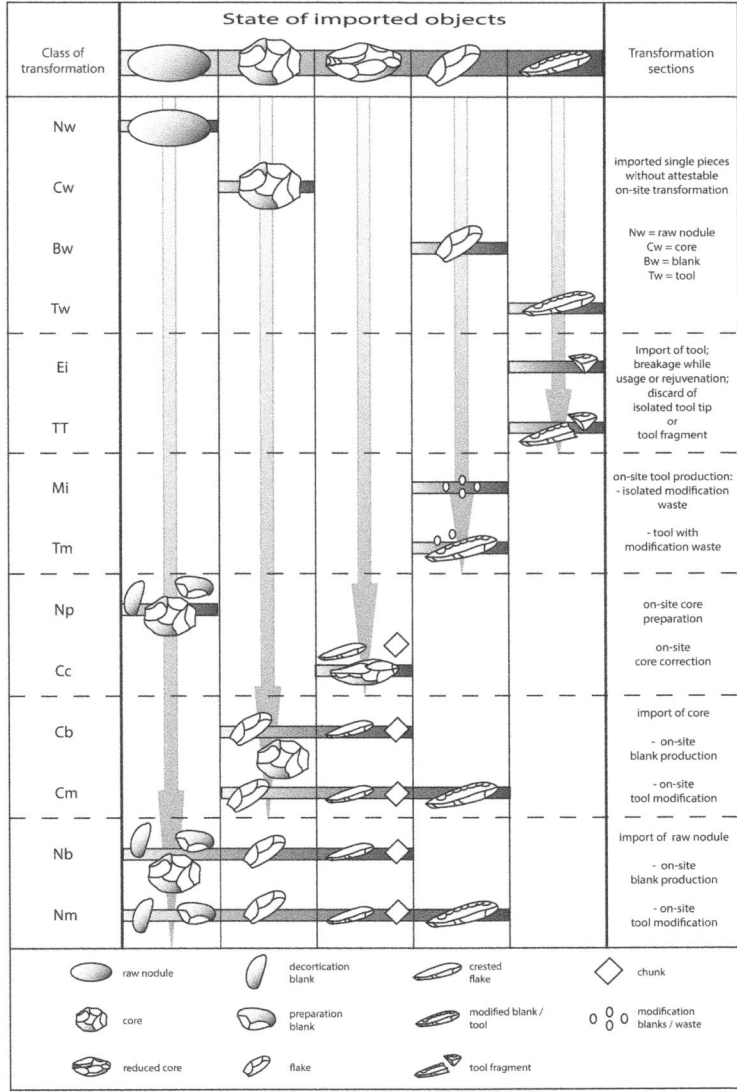

Fig. 3.5   Possible transformation sequences using the example of the nomenclature of transformation classes and transformation sections developed by W. Weißmüller (1995). (Graphic based on G. Bataille (2010: Fig. 2))

modified and transported, produced during occupation of the site are the result of deliberate manufacturing processes. These are distinguished from static categories of artefacts, which bear witness to certain manufacturing processes through their presence, but are themselves only by-products of these processes. Thus, only the presence of retouching chips can indicate the on-site modification of tools (Mi, Tm). The same applies to isolated tool fragments or defective tool fragments (e.g. broken tool tips), which indicate the use of equipment on site and the discard of fragments in the course of its damage (Ei, TT). Whether this code, developed on the basis of Middle Palaeolithic assemblages of the Sesselfelsgrotte, should be used can be left to the individual researcher. It remains important to note that by raw material sorting it is possible to identify events that took place at the site and to reconstruct related operational sequences. In this way is it possible to prove the affiliation of artefacts, which often cannot be clearly deduced due to taphonomic and other post-depositional processes, and to assign them to specific occupation events.

3.4   Examples of Use and Possible Interpretations

The TA ultimately serves to make interpretations about past activities in their functional and technological context. For example, the determination of core reduction concepts with regard to their technology is possible on the basis of a few characteristic objects belonging to a workpiece. For example, the TA has been used to demonstrate that specific lamellar microliths of a raw material unit were produced by means of the reduction of different burin core types (Bataille & Conard, 2018: Plate 5, 1–7) (Fig. 3.6). It is also possible to draw broader conclusions about occupation palimpsests (= multiple occupations) and related land-use strategies based on workpiece assortments and their spatial correlation with other find categories. For example, TA on artefacts of the Middle Palaeolithic stone assemblages of find horizons II/7e, II/8, III/1 and III/2 of Kabazi II (Crimean Peninsula) revealed that the individual raw material units were concentrated in different zones of the excavation area, which were associated with the remains of hunting spoils of different animal species in different activity zones. Raw material from different deposits could be correlated with these activity zones (Bataille, 2006, 2010, 2012: Figs. 5 and 8) (Fig. 3.7).

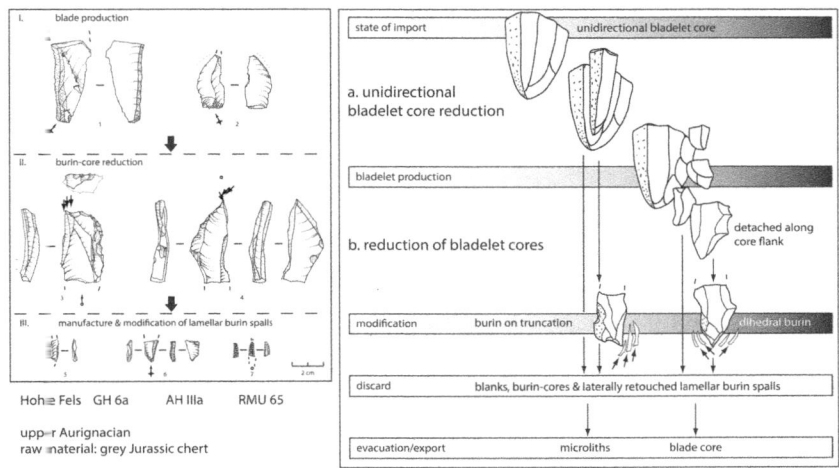

Fig. 3.6 Example of a raw material unit of the Aurignacian inventory AH IIIa of the Hohle Fels cave in the Swabian Jura (Germany). Few characteristic artefacts (left) can prove a specific manufacturing chain on a workpiece basis (right). (Graphic and artifact drawings: Guido Bataille, based on Bataille and Conard (2018: Appendix, Plate 5))

3.5   Possible Applications and Criticism

The TA offers the possibility to subdivide lithic assemblages into functional units. Based on this, statements can be made about applied technological strategies, single events in the context of multiple recurrent occupations and overlapping activity zones as well as taphonomic influences on archaeological horizons. This methodological approach thus offers manifold possibilities for the investigation of lithic assemblages. In this context, Weißmüller starts from the basic assumption that in the case of a pronounced heterogeneity in the raw material spectrum, artefacts can be assigned to raw material units with workpiece character, i.e. to original raw pieces. Whether this is actually the case can ultimately only be verified beyond doubt on the basis of refittings. However, since a complete refitting of lithic artifacts of an analytical unit is illusory due to missing intermediate pieces, refitting sequences within raw material units serve to validate employed sorting units. In addition, other methods can help to provide information about the quality of workpiece sorting. For example, high-resolution microscopic examinations can be used to minimize any uncertainties in raw material assignment. The examination of the variability and origin of used raw material as well as the study of regional raw

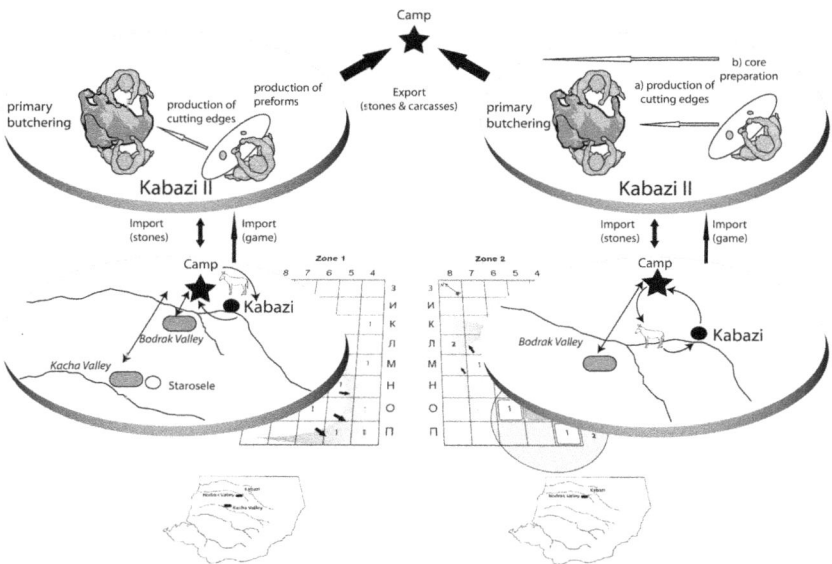

Fig. 3.7 Reconstructed land use system using the example of Middle Palaeolithic find horizons III/1 and III/2 from Kabazi II (Crimean Peninsula). The different variants of the land use system are based on raw material sortation and their spatial correlation with different activity zones at the site as well as the determination of the proportion of lithic material from different raw material sources in the individual workpieces. (Graphic from: Bataille (2012: Fig. 8))

material sources are essential components of the transformation analysis. In doing so, it may be possible to draw on existing regional studies of raw material occurrence, use, and exchange (e.g., Floss, 1994; Burkert, 1999; Çep et al., 2011). In general, to adequately describe technological transformation processes, the TA should always be combined with a comprehensive techno-typological attribute analysis. Furthermore, in order to minimize the room for interpretation of raw material sortings, a conscientious differentiation should be made between unique workpieces on a raw piece basis and common variants of an outcrop.

4.1   Introduction

Lithic technology refers to all activities of humans related to the production, transformation, and use of lithic objects (Inizan et al. 1999).

The *chaîne opératoire* approach is a **technological** analysis that uses each individual lithic object of an assemblage with its combined characteristics and thus, the assemblage in its entirety. The aim is to trace the logical sequence of the different stages of the **operational chain** (*chaîne opératoire*) from raw material procurement through blank production, tool manufacture, use and recycling to discard (see Leroi-Gourhan, 1964; Boëda et al., 1990; Geneste, 1991; Inizan et al., 1999; Soressi & Geneste, 2011). This **holistic** approach allows, on the one hand, to reconstruct the chronological **sequence** of the different steps involved, including production, transformation and use (Geneste, 1991). On the other hand, it is possible to understand the **spatial organisation of** the technological process (Geneste, 1985). The operational sequence can be inferred from the individual artefacts or their technical landmarks (i.e. type and position of negatives, abrasion or impact point). The presence or absence of the by-products of a technological stage, i.e. a concrete step within a specific reduction concept, allows to draw conclusions about the management of raw materials and/or end-products within a certain areal (Perlès, 1989).

The approach is descriptive and based on the fundamental principle that the **fracture mechanic** of conchoidally fracturing rocks is subject to three constraints, namely the mechanical properties of the rock, the geometric shape of the volume, and the application of force (Kerkhof & Müller-Beck, 1969; Dibble & Whittaker, 1981; Rezek et al., 2011; Porraz, et al. 2016). Once the knapper learns these fracture-mechanical constraints, they become fixed rules whose adherence allows for control and predictability of the fracture. Considered in its context, each lithic artefact represents the product of a particular **technological system** or, more precisely, its lithic sub-system. This sub-system interacts with other sub-systems, such as that of bone tools or wooden tools, within the larger technological system of the group to which its maker belongs in terms of intentions and knowledge (Inizan et al., 1999). The objective is to describe the different stages that led to the assemblage, in order to capture the relationships between the different production steps as well as characteristics specific to the manufacturing process. This makes it possible to subsequently understand their meanings in the overall cultural context (Porraz et al., 2016).

From a general point of view, the *chaîne opératoire* approach assumes that lithic production first emerges as a **cognitive project**, which is then translated, on an intellectual level, into a **conceptual schema**, which is finally concretised through a series of actions (operations), **an operative schema** (Fig. 4.1) (Pigeot, 1991; Inizan et al., 1999; Soressi & Geneste, 2011). All three steps are interdependent and can

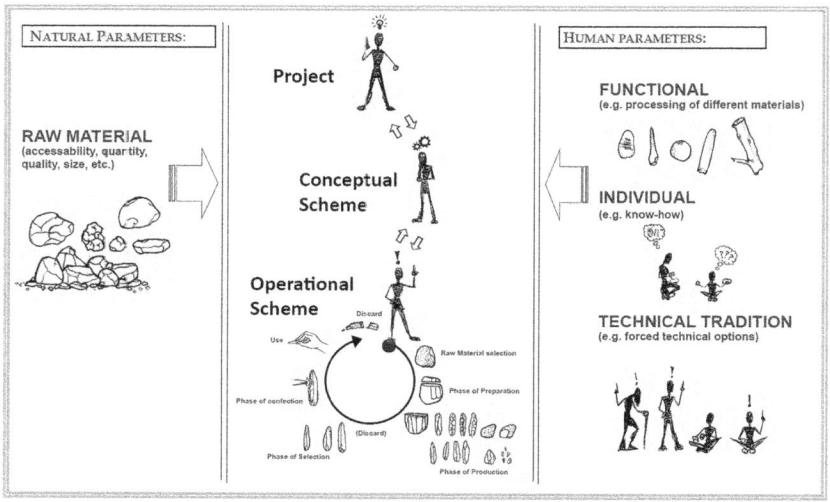

Fig. 4.1   Theoretical framework of the *chaîne opératoire* approach illustrating the relationship between cognitive project, conceptual schema, and operative schema. (Graphic: Viola C. Schmid, created on the basis of Soressi & Geneste, 2011: Fig. 3 & Porraz et al., 2016: Fig. 5; drawing by Heike Würschem)

be influenced by numerous, partially interacting, **natural** (such as raw material availability, quality and size) and **human parameters** (such as functional necessities individual know-how and technological traditions) (Boëda et al., 1990; Pigeot, 1991; Inizan et al., 1999; Soressi & Geneste, 2011; Porraz et al., 2016). According to the theoretical background, the observable constant and regular elements of the operational schema, i.e. more concrete reduction strategies (cf. Chap. 3), allow us to identify the underlying concept that drives the operative schema. Consequently, the inferred goals of the conceptual schema enable the determination of the original cognitive project. Thus, consistency or frequent recurrence of one or more patterns can be interpreted as intentional (Soressi & Geneste, 2011).

4.2 History of Research

The concept of technology as **science of human activities** was proposed in France by A. Leroi-Gourhan (Leroi-Gourhan, 1943), and later popularized by the science historian A.-G. Haudricourt (Haudricourt, 1964, 1987). Both were students of the French sociologist M. Mauss, who had earlier recognized the advantages of understanding a specific society through its techniques (Mauss, 1947). The term *chaîne opératoire* was first used by Leroi-Gourhan (1964, p. 164), who did not formalize it, but paved the way for its future use in ethnology and especially archaeology through his publications, his teaching at the Sorbonne University (later Paris I) and his leadership of the research team *'Ethnologie préhistorique'* (Audouze & Karlin, 2017).

From the late 1970s to the early 1990s, J. Tixier (Department *'Préhistoire et Technologie' of* the CNRS in Paris), M.-L. Inizan, H. Roche and their colleagues defended a new approach to prehistoric societies through the study of stone artefacts, which they qualified as a technological approach. In doing so, they went beyond the typological approach that had been commonly used until then to classify, thus advancing towards a deeper understanding of the **social meaning** of technological concepts used in the past and their respective modes of implementation (technique) (Tixier et al., 1980; Tixier, 2012). This approach shifted the focus from the study of prehistoric people through their stone tools to the study of **prehistoric societies** through their **cultural technologies**, which are understood not only as a social product but also as a founding element of the society (Schlanger, 1991). Consequently, the study of lithic technology provides insight into past communities in which technology emerged in different contexts.

Over time, Tixier and his colleagues introduced the concepts of technical system, different production processes and ways of technical implementation into archaeology (Soressi & Geneste, 2011). These fundamental principles had previously been formulated by French ethnographers working on material culture (cf.

Balfet, 1975; Cresswell, 1983). It is worth noting here that French researchers in ethnology and prehistory worked much closer together at that time than nowadays. In the late 1960s and during the 1970s, Parisian ethnologists held workshops on technology in which several prehistorians participated. The ethnologist R. Cresswell founded the research group *'Techniques et Culture' in* 1973 and the corresponding bulletin in 1976, which was widely read by prehistorians (Audouze et al., 2017). In archaeology, debate ensued as to how these technological concepts could be useful in describing and interpreting the variability observed in Palaeolithic industries in cultural terms. It was not until the 1990s that archaeologists around Tixier made their approach explicit; this is evident in the changes between the two main versions of their textbook *'Préhistoire de la Pierre Taillée'* (1995; Inizan et al., 1992; Tixier et al., 1980) and in works by their students and colleagues (see for example Boëda, 1986; Geneste, 1985; Pelegrin, 1995; Perlès, 1989). The 1995 version of their textbook (Inizan et al., 1995, 1999: 13 for the English translation) begins with a citation of Haudricourt that establishes the French technological approach: "While the same object can be studied from different viewpoints, that which consists in defining the laws of creation and of transformation of an object is undeniably the most essential of all viewpoints (Haudricourt, 1964 in Haudricourt, 1987, p. 38)." With this introduction, Tixier and his colleagues clearly highlighted one approach, that of the *chaîne opératoire* (Soressi & Geneste, 2011).

4.3   Procedure (Methodology)

Technique, Method and Concept

- **Technique** refers to the physical means of transmitted energy associated with the detachment of blanks. This includes, for example, knapping with or without an anvil, the shape and raw material of the hammer(s) used, the way the piece being worked is held, and other aspects of body technique (Tixier, 1967).
- The **method** refers to the intellectual steps followed during the process of reduction and materialized through the organisation of the negatives on the cores and blanks (Tixier, 1967).
- The **concept** describes the overarching theoretical framework to which the reduction methods are oriented. Thus, the concept (e.g. Levallois concept) may be maintained throughout the operational chain, while the method changes (e.g. from preferential Levallois (*méthode Levallois à éclat préférentiel*) to recurrent centripetal Levallois (*méthode Levallois récurrent centripète*) (Boëda, 1986).

The study protocol described here follows that of Soressi (2002) and Soressi and Geneste (2011). However, it should be adapted to the specific assemblage as needed. The technical instruments recommended are lamps, scales, calipers, goniometers, magnifying glasses and low-power stereo microscopes (cf. Chap. 7).

In a technological study, the first step in principle is to **separate the artefacts according to raw materials** (cf. Chap. 3). This is done on the basis of criteria that could have influenced the knapping process, such as the petrographic nature and the condition of the cortex. These aspects indicate the geological source from which the raw material was acquired and the context of the outcrop (primary, secondary, etc.). In addition, it is advantageous to group the lithic artefacts within the formed raw material units according to technological categories (cores, flakes, etc.).

The second step aims at gaining a thorough **understanding of the methods used** and their concrete implementation, the technique, which was used by the knappers. An exact determination of the technique for each individual piece is difficult, so an attempt should be undertaken to establish general trends on the basis of certain landmarks (e.g. lip, bulbar scar, Hertzian cone, impact point, type as well as form of the platform, platform depth, dorsal reduction and exterior platform angle). The study of reduction methods is of paramount importance and the organisation of negatives on each artefact should be considered here to reconstruct short reduction sequences. For this purpose, certain landmarks on the surfaces of the lithic objects are taken into account, which allow the determination of the chronology as well as the direction of the negatives to each other (cf. Chap. 5). By arranging these sequences in a sequential order, the global method(s) underlying the assemblage can be reconstructed. It is also helpful for this purpose to produce diacritic schemes (*schémas diacritiques*) or reduction schemes (cf. Chaps. 3 and 5), based on which a schematic representation of the operational chain can be generated, illustrating the overall sequence of the phases of reduction and production. Some of the lithic artefacts belong to technological categories, such as a primary crested blade, a core tablet, or a Levallois flake (for definitions of the terms see Hahn, 1991; Inizan et al., 1999), and are more informative than others in this respect because they originate from a specific step within the reduction process. Furthermore, some of these steps are so essential to the reduction sequence that their presence or absence is always significant.

The aim of the third step is to highlight the **morphological characteristics of the products of the operational chain**. These result from the techniques and methods used. "Mental refitting" (*remontage mental*) should guide all observations (Pelegrin, 1995). Moreover, the proposed model can at the end be tested and supported by physical refittings and experiments (Tixier, 1980; Geneste, 1991).

The final step is to determine if **each stage of the operational chain** is present in the assemblage **for each identified raw material unit**.

The observations, attributes and attribute combinations judged relevant during the classification phases should then be recorded and quantified in a database to allow the application of descriptive and comparative statistical tests. The definition of attributes is based on hypothetico-deductive reasoning and takes place *a posteriori*, which is one of the main practical differences between the *chaîne opératoire* approach and other technological approaches. Yet, some attributes already existing in the literature, some of which are proposed in Chap. 2, and/or further attributes that appear to be significant for answering certain technological, techno-functional or techno-economic aspects can be used (see Fig. 4.2).

4.4   Application Examples

After the study protocol has been successfully implemented with the inclusion of all blanks, cores and tools, it is necessary to capture the results of analysis and interpretation in a written and visual format. The two examples of application from the South African Middle Stone Age (MSA) on the *chaîne opératoire* approach are intended to illustrate, on the one hand, the reconstruction of the chronological sequence of the technological stages and, on the other hand, the spatial organisation of the operational chain.

The study of the lower MSA strata from Elands Bay Cave revealed that past people used almost exclusively slabs of quartzite without preparation as cores to produce mainly flakes. Only a few of these blanks were further modified. The flakes show different morphometric characteristics indicating the exploitation of the slabs in different planes of symmetry. The reconstructed operational chain is summarized in Fig. 4.3.

In the assemblage of the C-A strata of Sibhudu Cave, a differential economic management of various raw material units by past groups became apparent (Fig. 4.4). In the case of dolerite, the most abundant raw material, and hornfels, all the technological steps of the operational chain from decortication, blank production, tool manufacture, re-sharpening to discard took place on site and only a few pieces were exported. The toolmakers occasionally fell back on to the sandstone available directly at the site. The lithic remains demonstrate that use took place exclusively on the site. Quartzite and quartz however were exploited in the cave, but some of the tools were exported for use elsewhere. Isolated end-products and tools made from fine-grained silicate rocks (chert) were brought in and partly re-sharpened or discarded.

TMB III PBF

Unit: Altitude (z): ID: Décapage: Square:

Technology

RM: ☐ Sandstone ☐ Silexite ☐ _____
Fragmentation:
Patination:
Thermal Alteration:
Unmodified Surface:____/_____

Length:
Width:
Thickness:
Weight:
Unmod. Surf. Loc.:

Blank Type:
Profile: Sin: Dex:
Morphology:
Cross Section:

Techno-functional general information

Bilateral Symmetry:
Hierarchisation:
Discard Stage:

Fracture Type (distal):
Back:
Technique(s):

Polish:
Resharpening:
Recycling:

Tool

Active part:
TPA:
Tip Outline:
Section:
Macro:
Cross Section:

Confection:

Tip Angle:
Length:
Max. Width:
Max. Thickness:

Middle Part:
Cross Section:
Length   Width:
Confection:

	Sin.:	Dex.:
Del.:	Del.:	
Angle:	Angle:	
Section:	Section:	

Base:
Cross Section: Length:
Type: Max. Width:
Morphology: Max. Thickness:
Macro: Base Outline:
Confection:

Sin.:   Dex.:
Angle: Angle:
Section: Section:

Drawing

Remarks

*indicated measurements and angles taken on the photo

Fig. 4.2 Recording form with attributes specially compiled for the technological and techno-functional analysis of the bifacial pieces from the MSA site of Toumboura III (Senegal). (Graphic: Viola C. Schmid)

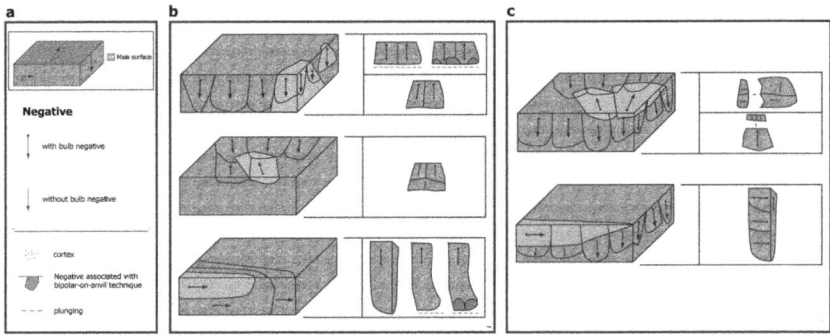

Fig. 4.3 Schematic representation of the operational chain in the lower MSA layers of Elands Bay Cave (Western Cape Province, South Africa): (**a**) The three planes of symmetry of a slab (1: Planar plane; 2: Orthogonal plane; 3: Linear plane) and the legend; (**b**) Scheme of independent orthogonal (top), planar (middle) and linear (bottom) slab reduction strategies; (**c**) Scheme of orthogonal and planar (top) and orthogonal and linear (bottom) combined slab reduction strategies. (Graphic: Viola C. Schmid)

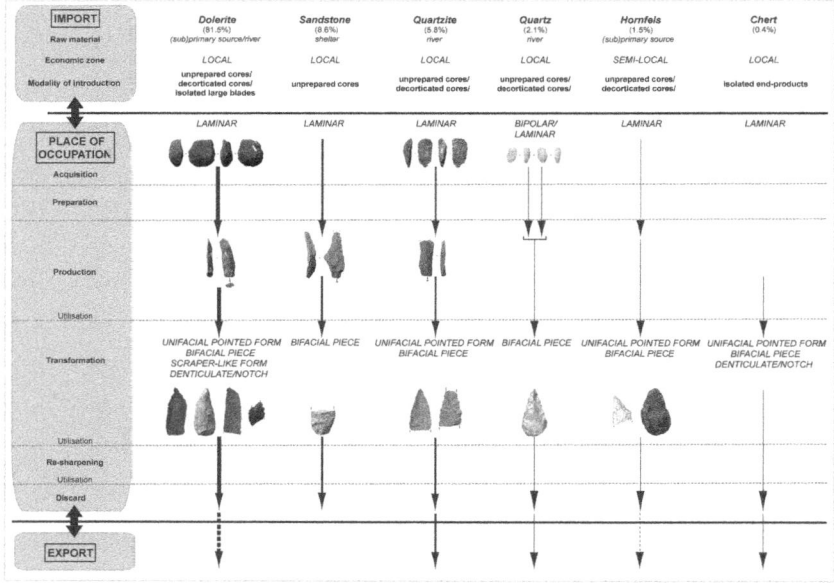

Fig. 4.4 Dynamic model of the techno-economic procedures concerning the different raw materials in the assemblage of the C-A layers of Sibhudu Cave (KwaZulu-Natal, South Africa). (Graphic: Viola C. Schmid, based on Porraz, 2005: Fig. 60)

4.5   Strengths and Weaknesses

The *chaîne opératoire* approach is predestined for a **targeted, pertinent answer to** questions concerning the technological knowledge and technical skills of past societies or the socio-economic and cultural context of knapping activities of a specific stone artefact assemblage. The dynamic approach envisages, after the initial appraisal of the stone artefacts, to *a posteriori* define, record, and quantify attributes deemed relevant. The advantage of this **precise attribute selection** is that the operational chain currently at hand can be addressed efficiently and flexibly. Since, in general, only phenomena that are comprehensible can be interpreted, and in the *chaîne opératoire* approach the attribute definition is based on an initial understanding of the assemblage, this technological analysis has a good degree of intersubjectivity if a similar approach is followed.

However, this approach requires more time investment, as it is necessary to familiarize oneself with the lithic material and to grasp its characteristics as best as possible. In addition, sufficient working space should be available to lay out the stone artefacts for the first overview. Ultimately, only assemblages with sufficient numbers of pieces will allow the various operational schemes to be accurately identified and documented. Finally, it should be noted that, for historical reasons in particular, the *chaîne opératoire* approach is characterised by a subjective wealth of experience and expert knowledge, and thus, the comparability between analysts is not unrestricted.

Working Stage Analysis

5

5.1 Introduction

In prehistoric archaeology we very rarely have the chance to grasp the actions of individuals. If we disregard human skeletal remains, the people living at that time are only visible through the objects of daily use left behind at the respective sites. The spatial distribution of artefacts or so-called features, such as hearths, can also provide information about the actions of people at individual sites. However, these finds and features at best reflect groups of people and not individually identifiable actors. With the help of the method presented below, the actions of individuals can be directly traced. In the German-Speaking research community, this method is known as **Arbeitsschrittanalyse (in Engl. Working Stage Analysis), and constitutes a realtively recent** analytical procedure. The basic assumption is that the artefact under investigation, be it a core for the production of flakes or a hand axe, is the result of a series of different work steps (Fig. 5.1). These working steps and the processing strategies behind them have been preserved on the surfaces of the stone artefacts in the form of negatives. The aim of the method is, on the one hand, to put these working steps, which can be derived from the negatives, into a **chronological order** and, on the other hand, to determine the function of the individual working steps. In the end, this results in the **reconstruction of the design process**, which at best reveals the individual handwriting of the maker.

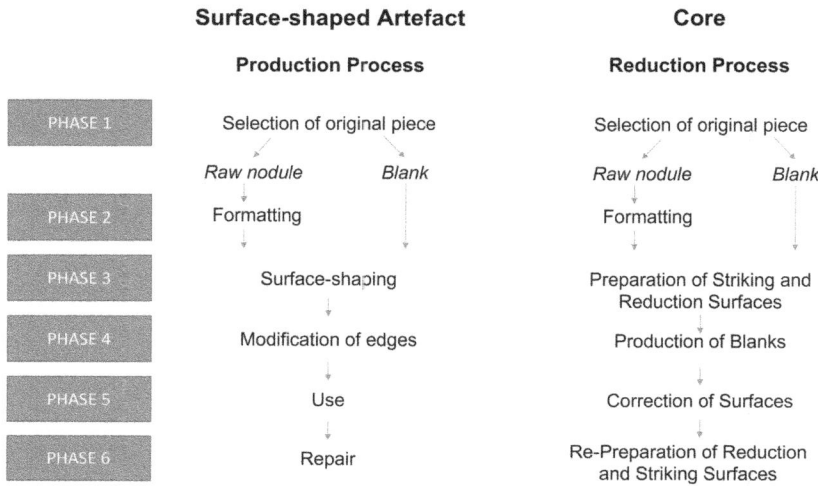

Fig. 5.1 Phases of the manufacturing process of surface-shaped tools (e.g., hand axes) and the reduction process of a core to produce blanks (e.g., flakes). (Graphic: Yvonne Tafelmaier)

5.2   History of Research

The scientific roots of the method presented here can be found in Francophone research, here above all in the so-called *schéma diacritique* (Dauvois, 1976). In order to obtain information about the genesis of an artefact, sign conventions have been introduced to code the different technological steps in the manufacturing process of a tool. Steps of different functions are marked with specific symbols so that information about the manufacturing process can be extracted from the drawing. Later, this type of representation was taken up, for example, by E. Boëda (*lecture des schémas diacritiques* 1986, 1988), J.-M. Geneste (1985) and L. Bourguignon (1992) and, moreover, often linked to the reconstruction of so-called operational sequences, *chaîne(s) opératoire(s),* which was also adapted for analyses on lithic technology at this time (cf. Chap. 4). In her PhD M. Soressi (2002) likewise reconstructed production processes by bringing single negatives on cores in a chronological order.

In Germany, the working stage analysis was formalized by J. Richter and A. Pastoors during their work on the material of the Sesselfelsgrotte, one of the most important Middle Palaeolithic sites in Central Europe. J. Richter described its main features in the context of his habilitation thesis on the upper Middle Palaeolithic horizons of that site, the G-layer complex (1997). However, the application of this method is missing in the publication. A detailed description can be found in an article published in French (Pastoors & Schäfer, 1999). Finally, the method found its first detailed application in the dissertation of A. Pastoors, which dealt with the Middle Palaeolithic lithic assemblage from Salzgitter-Lebenstedt in Lower Saxony (Pastoors, 2001). Manufacturing processes of cores as well as bifacially surface-shaped artefacts (e.g. hand axes, leaf-shaped scrapers) were reconstructed using the method presented here. O. Jöris (2001) was also one of the first to describe the production methods of so-called *Keilmesser* using a similar procedure. A summary in German can be found in J. Richter's (2018) published work "Altsteinzeit". In English, the main features of the method are presented in articles by Pastoors (2000a), Kurbjuhn (2005), Pastoors et al. (2015), and Bataille (2016, 2013), and by Pastoors and Schäfer (1999) and Richter (2001) in French.

Among the Anglo-Saxon research community, the work of Hassan (1988) is particularly noteworthy. He attempted to understand stone artefacts as end products of a cognitive process and pursued a theoretical approach entirely in the sense of and using terminology from N. Chomsky's *Generative Grammar* (1965). On this basis, he examined Late Palaeolithic artefacts from various sites in the Nile Valley and was able to highlight similarities in *competence* and *performance as* well as differences in *performance* in the production process.

5.3   How the Method Works

The workflow comprises the following steps: creating a technical drawing or a high-resolution photograph, determining and naming the work steps (coding) on the basis of the negative surfaces on the artefacts, recording the temporal relationships of adjacent work steps and determining their function. The subsequent creation of a Harris matrix serves the chronological reconstruction of the production process. These steps are explained in more detail below.

5.3.1   Which Artefacts Are Particularly Suitable?

In principle, the method can be carried out on all lithic artefacts whose surfaces preserve negatives of the manufacturing process. A single artefact forms an analytical unit. Particularly suitable are objects that have undergone a complex production process. On the one hand, this applies to surface-shaped artefacts such as hand axes, leaf points or asymmetrical bifacial backed knives (*Keilmesser*). But also cores, which are used for the production of blanks, with their often numerous reworked surfaces are adequate objects of study. Thus, on the one hand, the working stage analysis on these pieces can provide information on the underlying technological concepts and their implementation. At the same time, it can be made visible in which cases a technological implementation failed and a correction had to be made during the workflow.

> **Background Information**
> By **surface-shaped artefacts** we mean stone tools, whose **outline** and **cross-section** have been shaped by removals into the surface (Richter, 1997, p. 185). These include, for example, handaxes or leaf points. In conventional German terminology, these artifacts are referred to as core tools (*Kerngeräte*), on the assumption that unworked raw nodules or natural debris were used as raw piece. However, often larger flakes or, in specific cases, blades served as blanks for unifacially und bifacially surface-shaped artefacts. Furthermore, the term bifacial tools, which is also frequently used, ignores the fact that artefacts that have been surface-shaped on one side (unifacially) also exist. For this reason, at least some of our colleagues have adopted the somewhat broader term surface-shaping (*Formüberarbeitung*), based on the French term *façonnage* (Richter, 1997; Pastoors, 2001; Uthmeier, 2004c; Böhner, 2008; Tafelmaier, 2011; Bataille, 2017).

Definition of a Work Step

In the course of the process of creating an artefact, certain stages are passed through again and again (Fig. 5.1). The aim of the method is to combine as many negatives as possible into one work step and to represent the various stages of the process in an approximately correct manner. Negatives represent a **work step** if they can be traced back to a **common reduction step** and thus have the same function and were performed by the same edge in the same direction and technique. There must also be a direct chronological sequence of recognizable negatives. Remnants of cortex and natural fissures, as well as ventral surface remnants, are considered working steps that are not intentionally applied, but testify the selection of the blank.

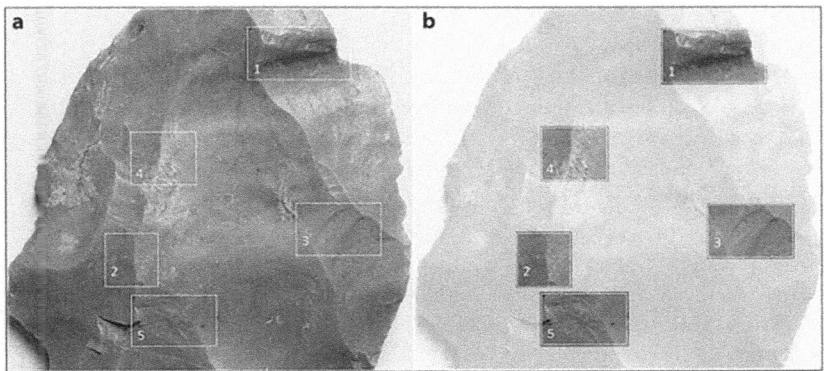

Fig. 5.2 Features defining the chronological sequence of contiguous negatives. (Graphic: Yvonne Tafelmaier, based on Pastoors et al., 2015)

5.3.2 The Chronological Sequence of Adjacent Negatives

There are different features on the surfaces which, either individually or in their entirety, allow statements to be made about the chronological sequence of negatives and thus work steps (Fig. 5.2) (Dauvois, 1976; Richter, 1997, p. 192; Pastoors et al., 2015, p. 67). They are sorted here in terms of their usefulness in practice and frequency of occurrence.

1. The younger negative cuts deeper into the raw material and shows a stronger concavity than the older one, especially at the common ridge.
2. At the common ridge the younger negative shows ray cracks/lancet cracks – those of the older one were overprinted when the younger one was removed.
3. The younger negative shows clearly pronounced Wallner lines in the terminal area.
4. Often there are splinters on the common ridge that accompany the lancet cracks of the younger negative.
5. The outline of the younger negative follows the relief of the preceding one.

5.3.3   Coding of Surfaces

Once an overview of the surface form and the work steps of an artefact has been obtained, the coding of the surfaces with regard to position, function and affiliation follows. There are different ways of assigning so-called "addresses" (Richter, 1997, p. 193). Standardized procedures have been proposed for surface-shaped artefacts (Richter, 1997; Pastoors, 2000a, b, 2001), Levallois cores (Pastoors, 2001), and laminar cores with multiple striking platforms and reduction surfaces (Tafelmaier, 2010; Bataille, 2016). However, apart from artefacts of these two categories, there are numerous other morphologically distinct forms or examples of use (Frick et al., 2017). One cannot do justice to this diversity with a standard procedure. Therefore, only a few basic guidelines will be presented here.

Regardless of whether it is a carinated core, a discoid core or a leaf point, for all artefacts it is important that a sensible orientation of the object is carried out before the addresses are assigned (Fig. 5.3). Subsequently, the assignment of addresses for edges and faces should be recorded in a sketch. For surface-shaped artefacts a subdivision into the upper side (O) and bottom side (U) is advisable at first. For trifacial artifacts, the third surface is also assigned a code (e.g., "R" for the back of a *Keilmesser*). For cores, reduction and striking surfaces receive their own designations. Sometimes surfaces act as both reduction and striking surfaces. In addition, cores may also have surfaces that have no function for the reduction process being described. However, it is important that all types of surfaces that can be distin-

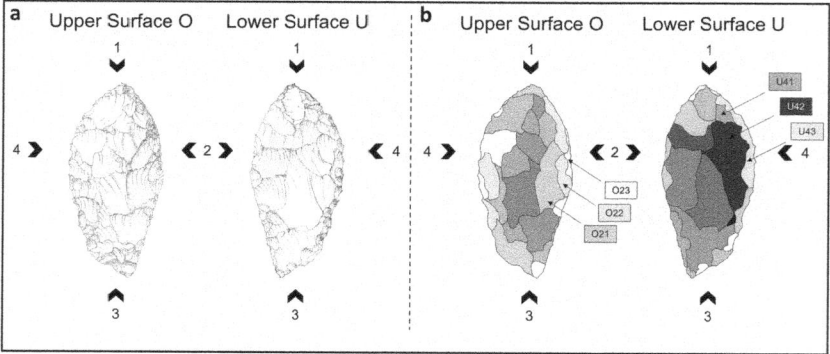

Fig. 5.3   Orientation of the artefact (a), coding and assignment of addresses (b) of the identified work steps. (Graphic: Yvonne Tafelmaier)

guished as individual areas by their location and/or function are coded as such. Here, too, a designation by letters is recommended.

If the surfaces are given letters, the edges (striking direction) are coded with Arabic numerals. In the case of a surface-shaped artefact, the tip (distal end) is given the number 1, and the other edges are numbered in a clockwise direction starting from this. On the underside, the numbering is done counterclockwise. Several steps found on one edge on the same surface are then numbered with consecutive numbers. For the coding, the chronological sequence is initially irrelevant and the consecutive numbering can be done arbitrarily. Accordingly, an address consists of at least three parts:

Example: Address **O21** (see Fig. 5.3)

1. **O** for the surface, top side
2. **2** for the edge and thus also the striking direction
3. **1** for step no. 1 on this surface and from this edge

Fissures, ventral or cortical surfaces have no relation to an edge and are therefore marked with the number 0. They usually represent the oldest recognizable surfaces on the artefacts.

5.3.4   Classification of the Working Steps with Regard to Function

Work steps are sequences of negatives with the same function. A very important basis of the method is to combine as many negatives as possible into one working step. Figure 5.1 shows the sequence of different working steps in the manufacturing and dismantling process of surface-shaped artefacts and cores. In the following, some fundamentally observable functions of different negatives will be explained. For all categories, the following applies: Remains of ventral or cortex surfaces testify to the selection, respectively provision of the original raw piece. No function intentionally constructed by the stone knapper is assigned to them.

In surface-shaped artefacts, we distinguish between **plane** and **convex surface shaping** (Boëda, 1994). Whereas in the former the removals often end in hinges and produce plane surfaces, in the latter convex surfaces are produced. These two types of surface-shaping were variously combined by Pleistocene hunter-gatherers. Boëda (1994) derived five different concepts from them: biplan, biconvex, plano-convex, plano-convex-plano-convex, and convex-plano-convex. The different concepts can be identified, apart from the cross-sections on the artefacts, by a different chronological sequence of the working steps. For example, in the case of plano-convex designed tools, a flat shaping of the lower side is carried out first, followed by a convex reworking of the upper side, which uses the lower side as striking surface. In the case of biconvex forms, shaping can take place alternately, i.e. the top and bottom sides are used alternately as striking surfaces.

The final phase in production is edge retouching. In this process, small retouching negatives close to the edge are used to construct a working edge, which offers different functional possibilities depending on the type of modification (Hahn, 1991, p. 169 ff.). In addition, artefacts often show chipping near the edge, which is distinguished from intentionally applied modifications by a less regular juxtaposition. These can be caused by use or by sediment retouching during embedding and storage in the sediment or by edge damage during recovery of the artefacts or find processing (Hahn, 1991, p. 167). These types of modifications are also recorded in the reconstruction process.

As far as the analysis of cores is concerned, attention to the different phases of reduction is of great importance. What all blank production concepts have in common is the creation of convex surfaces, which enables the targeted extraction of desired shapes. In many cases, the first stages of reduction, i.e. the initialization of raw nodules, are no longer recognizable. However, the establishment and preparation of the striking surfaces as well as the technological concept, i.e. the preparation of the reduction surfaces, can usually be traced. In addition, the lateral and distal convexity for the reduction of the target products can be created in various ways. For cores that have not been reduced according to defined concepts, such as the Levallois concept or the discoid concept, the coding can be handled more flexibly due to the lack of standardization.

5.3.5 The Reconstruction of the Temporal Sequence Using a Harris Matrix

Once the work steps have been coded, the function determined and all the information recorded in a database, the determination of the chronological sequence can begin. The goal is to create a Harris matrix that, similar to the stratigraphic sequence of an archaeological excavation, maps the work steps in the chronologically correct order (Fig. 5.4). In the best possible case, a work step can be found on each rank position and the entire chain of action can be clearly reconstructed. However, this is only likely in very few cases, because in most cases it is not possible to relate all the work steps to each other in terms of their temporal relationships.

Fig. 5.4 Example of a Harris matrix showing the chronological sequence of the defined work steps (Software: ArchEd)

With the help of the characteristics explained above (Fig. 5.2), the temporal relationships between adjacent work steps are determined. For example, for step O22 (Fig. 5.3): O22 is younger than O21, but older than O23. All identifiable relationships are recorded in a table. The following columns should be included: "younger than", "older than", "at the same time as". Since especially artefacts that are used for a long time and are characterized by extensive revision and a high number of work steps and many temporal references, the calculation and evaluation can only be done with the help of software. These are available as freeware or as commercial programs. A good free solution is for example the program Stratify 1.5 (Herzog, 2010). The work steps are entered into the selected program as stratigraphic units with a classification of their chronological relations. On this basis, the software calculates a chronological sequence of the documented work steps and checks the entered data for circular reasoning (Fig. 5.4). Basically, the program sorts work steps whose rank position cannot be determined unambiguously into either the most recent or the oldest possible position. It is up to the user to decide which of these two options is implemented. It is recommended to choose the first option, because especially the oldest phases of the manufacturing process provide important information about the conception of an artefact and should be recorded as clearly as possible. Finally, the Harris matrix developed must be checked for consistency against the original observations.

5.4   Example: Reconstruction of the Manufacturing Process Using the Example of a Middle Palaeolithic Leaf Point

Once the entire documentation process (sketching, coding, recording the chronology, creating the Harris matrix) has been successfully carried out, it is necessary to write down and visualise the interpretation of the analysed artefact and thus present it in a way that can be understood by an external observer. Figure 5.5 shows a working stage analysis of a Middle Palaeolithic leaf point. The technical drawing in the upper part gives an impression of the shape of the examined artefact. Colouring of different working stages of varying function then provides an initial visual record of the manufacturing process. Darker areas indicate early stages of the production process, lighter colored areas represent late stages. For clarity, the Harris matrix at the bottom of the figure has been combined with the function assigned to the working steps. In this way, the biography of the artefact can be comprehensively reconstructed.

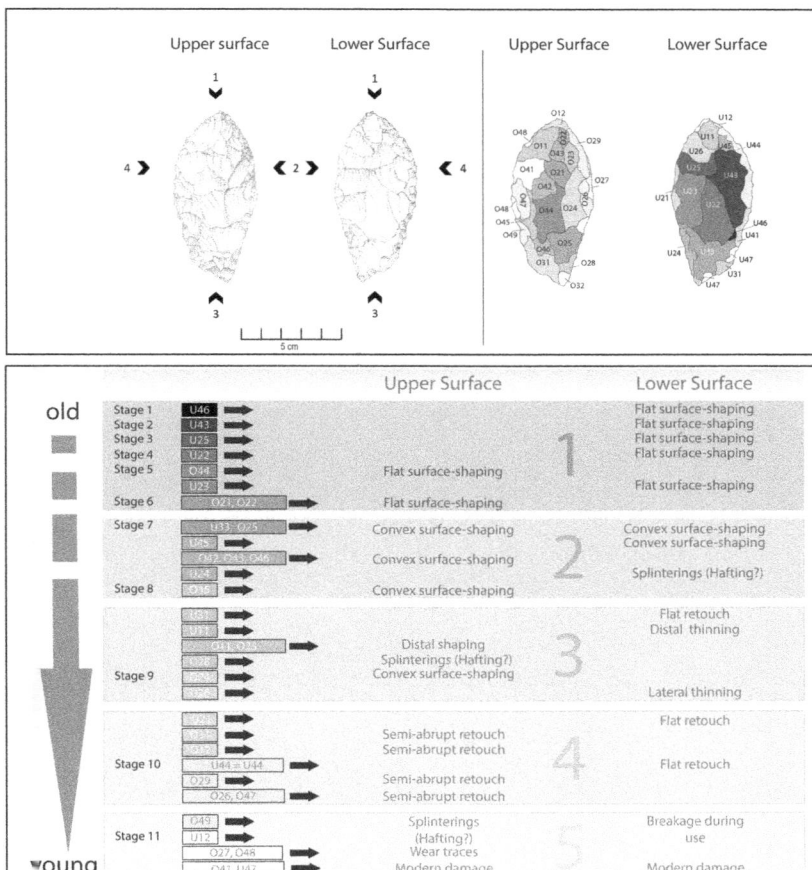

Fig. 5.5 Reconstruction of the manufacturing process of a Middle Palaeolithic leaf point. (Graphic: Yvonne Tafelmaier)

5.5   Strengths and Weaknesses

Advantages: The method offers the possibility to directly reconstruct action and decision processes of individuals. A few artefacts are sufficient to understand aspects of cognitive processes of hunter-gatherers. An analytical unit consists of only one artefact. The amount of material required is low and the standardized analysis procedure can be learned comparatively easily.

Disadvantages: Performing the analysis is time-consuming. It takes significantly more time to examine an artefact compared to attribute analysis (Chap. 2). In particular, the technological classification of the individual work steps is criticized as being very subjective (Bar-Yosef & van Peer, 2009).

Analysis of Techno-Functional Units

6

6.1 Introduction

The **design** of a human-made tool is in most cases an interplay of various factors, whereby, in addition to aesthetic preferences, the **functionality** of the object plays a role. The method presented here aims to reconstruct the function, **mode of operation** and **use** of concrete lithic artefacts. For this purpose, mainly those artefacts are selected which have intentionally modified edges, i.e. which represent tools in the analytical sense. It is closely related to working stage analysis (Chap. 5), which traces the manufacturing process dictated by the design, and microscopic analysis (Chap. 7), which can provide information about the use of a tool. What all three methods have in common is that they deal analytically with only a single object. If all three methods are combined, a maximum of information can be obtained and the results can be checked against each other.

6.2 History of Research

The method of analyzing so-called **techno-functional units** *(unités techno-fonctionelles)* originates from the French school, which is considered a pioneer with its early focus on technological issues. It was Michel Lepot, in his never-published 1993 thesis, who first introduced the concept of techno-functional units into scientific discourse and developed the basic terminology. Building on this and influenced by the work of the psychologist P. Rabardel (e.g. Rabardel, 1995), who dealt with the interaction of human and tool and ergonomic as well as cognitive aspects, E. Boëda (2001, 2013) applied the method systematically and comprehensively and extended its theoretical foundation (Boëda, 1997). Meanwhile, the method is used in the context of numerous studies and is not confined to specific regions or to selected epochs. While the first works deal with assemblages and artefacts from the Palaeolithic (Boëda, 2001; Pastoors, 2001; Soriano, 2001) there are also examples of applications from the Neolithic (e.g. Donnart, 2010).

6.3   Methodological Principles

If we look at a commercially available knife, it always consists of at least **three functionally different components,** which only guarantee functionality through their interconnection (Fig. 6.1). The knife has a handle that is held by the user, a blade that comes into contact with the object to be worked on, and a body that connects these areas. A **tool unit** thus has an averaging function and transfers energy from the user to the object to be worked on in a predictable way. The method of analysis of techno-functional units (*unités techno-fonctionelles/UTFs*) developed by M. Lepot (1993) makes use of this fact. As explained by the knife example, a **tool unit** consists of **three techno-functional units** (Fig. 6.1). An area where the energy is absorbed – in our example the knife handle. This area is called *contact*

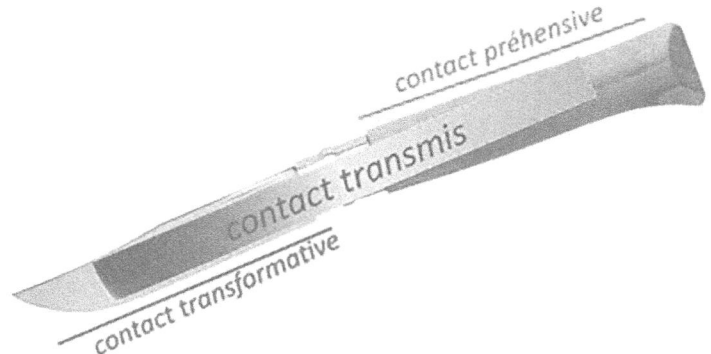

Fig. 6.1 A knife has different techno-functional units: (1) a handle, the passive part (*contact préhensive*) at which the energy arrives, (2) a mediating part (*contact transmis*), the overall plastic body that transmits the energy, and (3) a sharp edge, the active part (*contact transformative*) that transfers the incoming energy to the object being worked on. (Graphic: Yvonne Tafelmaier)

préhensive (**passive part**). Secondly, a working edge that transfers the energy to the object to be worked. This component is called *contact transformative* (**active part**). Finally, the plastic body mediating between the two functional areas, which grants the flow of energy from one component to the other. It is called *contact transmis*. This last component is often difficult to separate exactly from the functional areas close to the edge – often the *contact transmis* forms a unit with the passive part (Boëda, 2001, p. 53). In practice, capturing this component consists of describing the shape of the blank (e.g. cross-section) and the underlying production concept (surface-shaped blank or leaf point). For example, in a Paleolithic asymmetric backed knife (*Keilmesser*), the mediating part could be designed as a bifacially surface-shaped blank with a plano-convex cross-section. The demarcation to the two functional edge areas (active and passive) would be possible on the one hand via the edge retouch (active part) and the back opposite the working edge (passive part). Unlike the bifacially surface-shaped blank with negatives of plane and convex shaping, the passive part can consist of a natural back.

The synergy effects between these three components condition the functionality and handling of a tool (Boëda, 1997, p. 34). A tool thus combines in its conception the constraints imposed by **instrumentalization** (design) on the one hand and **instrumentation** (handling/use) on the other (Boëda, 2001, p. 52) (Fig. 6.2). Unlike in our simple knife example, an object can also unite several *tool units* (i.e. *tools*) in itself. For example, a lighter can be used to light a candle; the passive part *(contact préhensive)* is held in the hand, whereas the active part *(contact transformative)* lights the fire. Very often the lighter is also used as a bottle opener. In this case

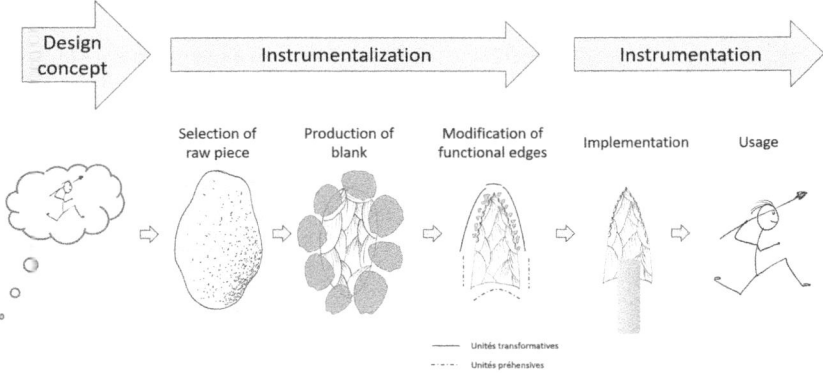

Fig. 6.2 Production Process of a stone tool from design idea, to technical realization (*instrumentalization*), to use (*instrumentation*). (Graphic: Yvonne Tafelmaier)

the active part becomes the passive and vice versa. The base of the handle now becomes the actual working edge and comes into contact with the object to be removed, the crown cap, and detaches it from the neck of the bottle.

Among Prehistoric lithic tools, the combination of several tool units within one object is the rule rather than the exception. Redesign during the period of use of a tool must also be considered. Considering that stone artefacts have been repeatedly reworked and used over longer periods of time, it becomes clear that the analytical recording of the individual techno-functional units as well as the tool units reconstructed from them represents a great challenge for the scientist.

6.4   Implementation of the Method

The implementation of the method is based on two essential components: on the one hand, technological observations of the artefacts are recorded in a sketch or drawing of the object under consideration in a standardised manner. On the other hand, qualitative (e.g. nature of the blank, outline and cross-section) and quantitative data (e.g. edge angle) are recorded and stored in a data table.

As with the working stage analysis, the analysis of techno-functional units is concerned with the reconstruction of the individual decision-making processes of prehistoric people with regard to the design of a tool. The ideal conception of the tool in the mind of the maker determines the selection of the material, the shaping into the desired initial form, the creation and reshaping of the edges, if necessary the fitting into a shaft *(hafting)* and finally the use. The task of the analyst is therefore to identify different functional areas, the so-called techno-functional units, on a stone tool. As far as the distinction between active and passive parts is concerned, all those sections of an edge belong to a **functional area** that coherently show the same **edge outline** and **edge angle** and are either uniformly retouched or not retouched.

The observation that edges of lithic artefacts are quite different in nature is a fundamental condition of the method (Wilmsen, 1968). In order to fulfil a certain function, the functional edge, as the active part of a tool is often called, must have a certain edge angle and a certain edge profile. The edge angle is the angle created by the adjoining lower and upper surfaces. In order to obtain the desired shape, an edge that can be reworked must first be created. This is done by producing a blank, e.g. a blade or a flake, or by providing a surface-shaped blank with edges that are still unmodified. By intentional modification of an edge *(retouching)*, its shape can be changed. If, for example, an edge is straight in plan view and has an edge angle of <45°, it certainly fulfils a different function than an edge that has an angle tending towards 90°. In the first case, the edge is sharp, in the second case, it is blunt (not to be confused with obtuse angle!). In very simplified terms, cutting operations can therefore only be carried out with the first edge. There are also cases where natural areas that already have the desired shape are included in a functional edge. This can be the case, for example, with a very blunt edge. A so-called natural back is thus included in the shape of the tool.

As for the determination of the edge angle, there are different methods of measurement. Control experiments on the accuracy and reproducibility of different measurement methods, favor the use of a caliper to measure the thickness of the edge area at a fixed distance from the actual edge (Dibble & Bernard, 1980). Then, the formula $? = 2\left[\tan^{?1}\left(\dfrac{5T}{D}\right)\right]$ is used to calculate the edge angle, where T

is the measured thickness and D is the distance from the apex to the measurement point (Dibble & Bernard, 1980). In the author's opinion, measurement by protractor (goniometer) has also proved useful in practice. Even if a certain fluctuation of about ±5° deviation must be taken into account, the measurements provide meaningful and reliable data. Care must be taken to ensure that the measurement is taken at a short distance from the edge. Our test objects are not industrially manufactured pieces, so that the measurement points must be individually adapted to each artefact. Further, the edge angle at different locations of a techno-functional unit cannot be expected to be the same to within two degrees. However, it should be within a certain range for a techno-functional unit. For example, it makes sense to form groups, that is, to distinguish between edges with angles up to 45°, between 45° and 70°, and between 70° and 90°.

The techno-functional units are usually designated with capital letters (A, B, C, etc.) (Fig. 6.3). It should be further noted that the top and bottom areas of an artefact within a functional area may be both passive and active parts of different tool units. A former tool edge of a dorsally laterally retouched blade may at a later stage be the passive part of a tool with, for example, a corresponding ventrally reshaped edge. The upper and lower sides of these areas are then additionally marked with numbers (A1 & A2), but make up a techno-functional unit. It is recommended to first determine the active techno-functional units *(contacts transformatives)* of an artefact. Regularly retouched edges that have angles less than 45° are often (but not always) identifiable as active components. Dorsally blunted areas, such as on back knives, are indicative of passive tool components. Adherent glue residue in this area could be further evidence of the passive nature of this area and suggest hafting of this side. Missing edge modifications as well as irregular chipping at edges can indicate both passive and active tool parts.

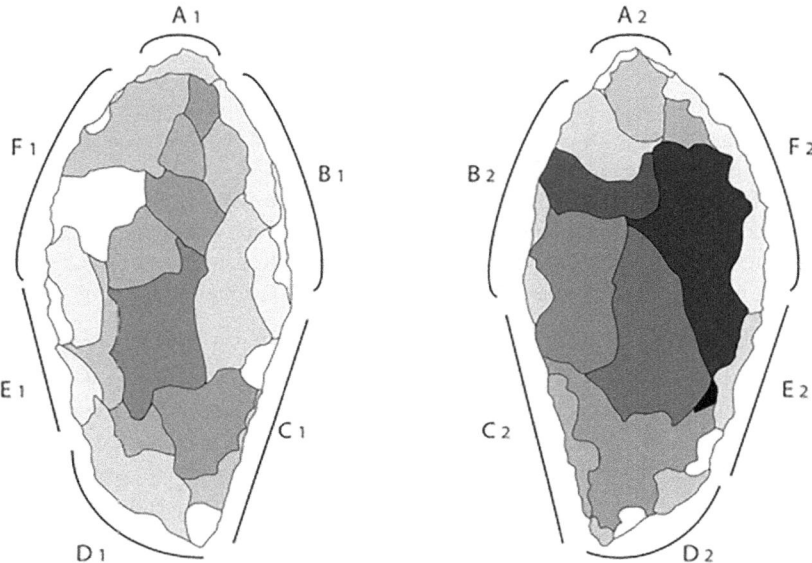

Fig. 6.3 Determination of techno-functional units and coding of artefact areas. Indispensable is the accompanying creation of a data table characterising the recorded techno-functional units of an artefact. (Graphic: Yvonne Tafelmaier)

The final reconstruction of a tool unit is based on the postulate of technical coherence, i.e. the meaningfulness of the combination of identified techno-functional units. Usually, the active components are directly opposite to the passive parts. For example, if the analysed tool is a drill with a shaped tip, the passive part *(contact préhensiv)* is located in the opposite tool area. The example given in Fig. 6.4 shows the combination of different active and passive parts within a single lithic artefact as well as the reconstruction of different tool units.

Workflow

- Working material: magnifying glass, microscope if necessary, drawing/ sketch of the artefact, protractor, pencil and coloured pens, data sheet (preferably digital)

1. Determination of the blank, its conception and shape, such as outline & cross section(s) = > Entering observations by characteristics in data table.

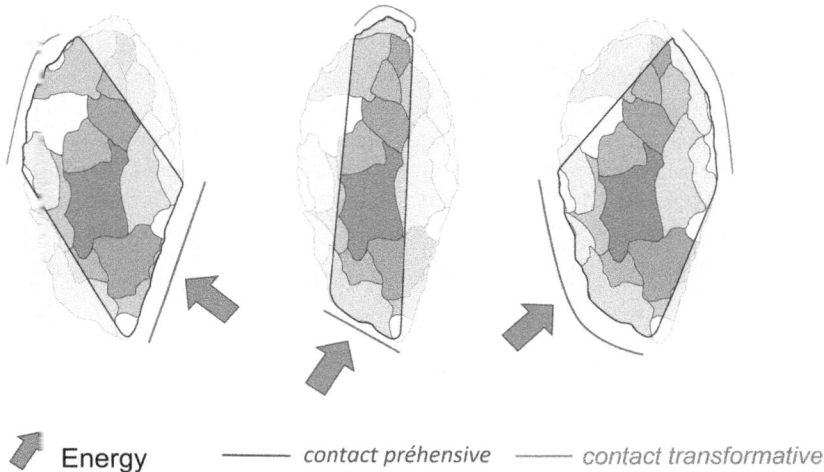

Energy ⟶ *contact préhensive* ⟶ *contact transformative*

Fig. 5.4 Functional reconstruction of a Middle Palaeolithic leaf point. The figure shows on the one hand the techno-functional units and on the other hand the three different tool units based on different combinations of the identified techno-functional units (*blue line*: passive tool area; *red line*: active tool area; *arrow*: shows the direction of energy transfer). (Graphic: Yvonne Tafelmaier)

2. Analysis of the areas near the edges: Position and shape of retouching (incl. edge angle), edge changes (e.g. use wear/cryo retouching, impact fractures) = > Drawing documentation and entering the measured values in data table.
3. Recording the techno-functional units = > Naming and graphic documentation in the sketch
4. Combination of techno-functional units to tool units (consisting of active – mediating – passive part) = > Virtual implementation of the interpretation

6.5 Criticism

The analysis of techno-functional units and the grouping into tool units is based on an interpretation of the synergetic interaction of different tool areas. In particular, the reworking and overlapping of different functional areas represents an uncertainty factor in the reconstruction of the design. A chronological sequence can often not be proven with certainty (cf. Chap. 5). While the recording of a techno-functional unit succeeds more reliably, especially the grouping into tool units is interpretative and therefore has a higher degree of subjectivity. Even E. Boëda, an advocate of this approach, notes that even if an interpretation is technically coherent, other possible combinations of functional areas can be valid in principle (Boëda, 2001, p. 62). The reconstructed tool units are thus to be regarded as hypotheses that should be tested by further investigations, such as analyses of wear traces (cf. Chap. 7).

7.1 Introduction

The exploration of wear traces preserved on stone artefacts, which can provide clues to the function and use of these pieces, is of central importance for the examination of Stone Age lithic assemblages. Based on evidence of the materials worked and the actions performed, prehistoric people behind the artefacts become visible with their potential intentions, behavior and knowledge. In addition, such investigations make it possible to document the working of materials that are usually no longer preserved (e.g. wood, plant fibres, hides/skins/leather). In some cases, analyses might be able to microscopically detect adhering remains on stone artefacts. In addition, microscopic analyses of lithic objects provide information about technologies or technological systems (e.g. about methods of hafting or the use of adhesives). The interest of use-wear analyses is the investigation of the concrete working method of a tool and the worked material. Use-wear analysis is an essential method of prehistoric archaeology, which brings to life the knowledge of the everyday organisation of Stone Age societies (Marreiros et al., 2014).

7.2 History of Research

Any use of stone artefacts can lead to the formation of traces of use on the surface when working with the piece itself, or even through hafting of the implement (cf. Rots, 2010; Keeley, 1982). Although, for example, the naked-eye sickle-gloss ("Sichelglanz") often observed on Neolithic flint inserts has been known for a relatively long time (Spurrell, 1884, cited in Pawlik, 1995). In terms of research history the first tangible beginning of use-trace investigations can be dated to 1957, when the original Russian-language edition of S. A. Semenov's standard work *"Prehistoric Technology – an Experimental Study of the oldest Tools and Artefacts from traces of Manufacture and Wear"* (title of the English translation by M. W. Thompson from 1964) was published. Subsequently, Western scholars soon turned to the subject (e.g. Nance, 1971; Keeley, 1974, 1980; Hayden, 1979; Moss & Newcomer, 1982; Moss, 1983; Plisson, 1985; Unrath et al., 1986). As use-wear studies became more widespread, more work was done to place interpretation on a firm experimental basis, including replication studies (e.g. Keeley & Newcomer, 1977; Keeley, 1980; Odell & Odell-Vereecken, 1980; Moss & Newcomer, 1982; Fischer et al., 1984; Plisson, 1985; Vaughan, 1985; Unrath et al., 1986; Beyries, 1993, 1999; Pawlik, 1995; Caspar & de Bie, 1996; Hardy & Garufi, 1998; Rots, 2010; Pétillon et al., 2011).

7.3   Methods

Prior to microscopic analysis, the stone artefacts must be cleaned from adhering sediment, and the pieces should be examined for any residues before cleaning (Pawlik, 1995; Rots, 2010); if necessary, these residues should be documented for subsequent analysis. Ensuing cleaning should be kept as short as possible, to prevent the formation of artificial traces and to avoid damaging any residues, and should be carried out with low-dose acidic solutions at most, so as not to affect potentially existing polishes or even remove them (Keeley, 1980; Pawlik, 1995; Rots, 2010). Cleaning the artefacts in an ultrasonic bath, possibly with the addition of some dishwashing detergent or a 1% potassium hydroxide solution is also possible (Pawlik, 1995). During the ongoing investigations the artefacts are cleaned either with alcohol or acetone. Both do not affect the traces of use, but are very suitable for removing adhering grease (from the fingers) or residues of adhesive clay (for fixing artefacts) (Rots, 2010).

Microscopic analysis of residues examine remnants of a material formerly attached to or worked with the stone artefact (Fig. 7.1; cf. e.g. Briuer, 1976; Anderson, 1980; Anderson-Gerfaud, 1986; Lombard & Wadley, 2007; Rots et al., 2015; Yates et al., 2015). In addition to identifying, for example, wood cells (e.g., Hardy & Garufi, 1998) or blood (Loy, 1993), insights into the chemical composition of residues (e.g., adhesives) can be obtained using various methods such as infrared spectroscopy (e.g., Baales et al., 2017), Raman spectroscopy (e.g., Bradtmöller et al., 2016), or gas chromatography and mass spectrometry (cf. Perrault et al., 2016; Cnuts et al., 2018).

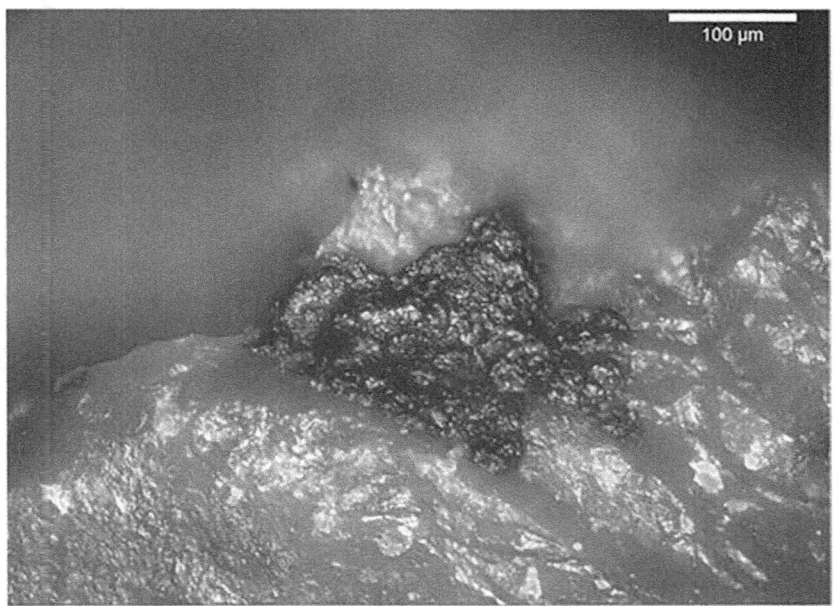

100 µm

Fig. 7.1 Microscopic detail of an adherent material residue on a lithic artefact. (Photo: Andreas Taller)

The signs of use on lithic artefacts can manifest themselves in the form of **polishes** and demarcated, **polished areas** *(bright spots)* on the artefact surface, **rounded edges** and **burrs, chipping** at the working edge and **fractures of** the artefact, as well as so-called *striae* (grooves deepened into the surface, which are formed by contact with relatively hard material and are often linear in shape, indicating a direction of movement). When used as projectiles, stone artefacts sometimes show characteristic fracture patterns that, when microscopically visible traces (e.g. *striae*) are taken into account, can prove such use (e.g. Fischer et al., 1984; Pétillon et al., 2011; Rots & Plisson, 2014).

The individual forms of wear traces must be integrated into an overall picture consisting of all macro- as well as microscopic traces, artefact morphology and archaeological context in order to enable a conclusive interpretation.

In microscopic examinations, a distinction is made between *low power* and *high power* **analyses**.

Low power analyses are performed with stereo microscopes (magnification up to 100×), which do not have their own light source in the objective. Stereomicroscopy follows the incident light principle, with an external illumination shining in from the side (Fig. 7.2a). The advantage is that despite the relatively low magnification, a three-dimensional image of the sections under consideration is obtained with good depth of field; this allows changes at the edges and in the surface topography of an artefact to be detected (cf. Pawlik, 1995).

For *high power analyses*, reflected light microscopes with possible magnifications of up to 1000× are used, mainly with objectives or combinations of 200× (e.g. Odell & Odell-Vereecken, 1980; Rots, 2010). These devices are also referred to as "metallurgical microscopes", following their frequent use in industry for surface inspection and material analyses (Fig. 7.2b). The object to be examined is again illuminated from above, whereby the light source is usually located in the objective itself.

Another possibility for *high-power* analyses is *scanning electron* microscopy (SEM; cf. Ollé & Verges, 2014). Here, the object to be examined is scanned by an electron beam in a predefined grid, and a high-resolution surface image with a large depth of field is generated from this information.

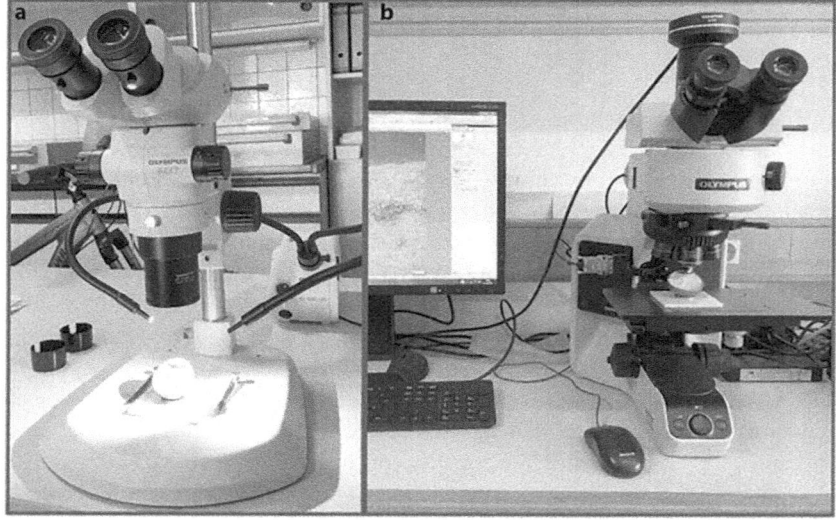

Fig. 7.2 Stereomicroscope (**a**), metallurgical reflected-light microscope (**b**). (Photo: Andreas Taller)

High power analyses are time-consuming, as the complete artefact surface is examined in detail for changes; informative documentation of any polishing or rounding is possible due to the high magnifications that can be achieved (Fig. 7.3).

All observations are continuously documented photographically and recorded on technical drawings or artefact photographs (Fig. 7.4); the microscopically obtained findings on the use of a stone implement are localised and interpreted in a comprehensive synopsis.

The *low power* approach can be used in addition to the detailed examination with a metallurgical reflected light microscope (e.g. in the sense of an enlarged sample, as this method is less detailed but can be carried out more quickly), but also as preparation for the *high power* examination (pre-selection of pieces).

The classification of the observed traces is only possible through comparison with a reference collection. Therefore, the creation, evaluation and archiving of comprehensive experimental series (i.e. collections) is an absolutely necessary pre-

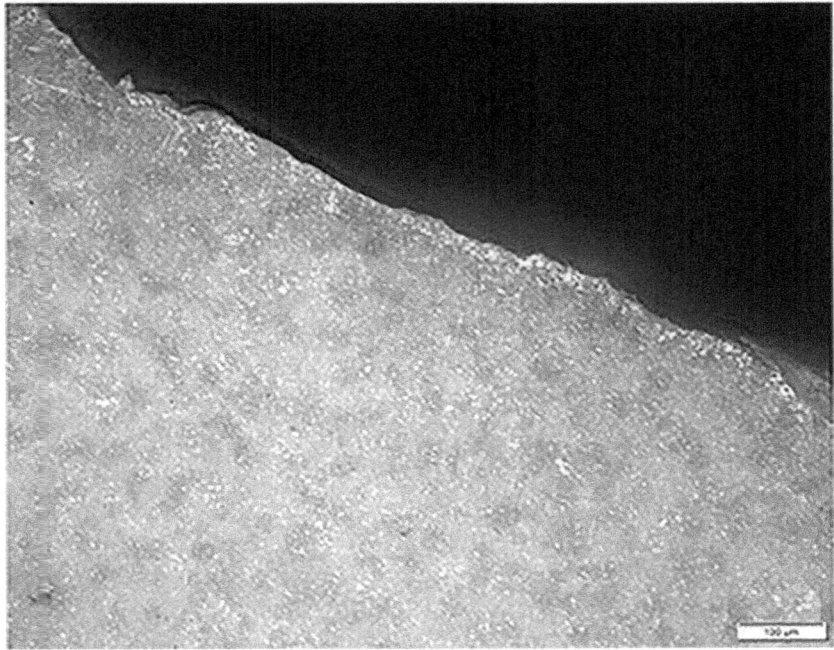

Fig. 7.3 Rounded, faintly polished edge of a lithic artefact used to work reindeer antler. (Photo: Andreas Taller)

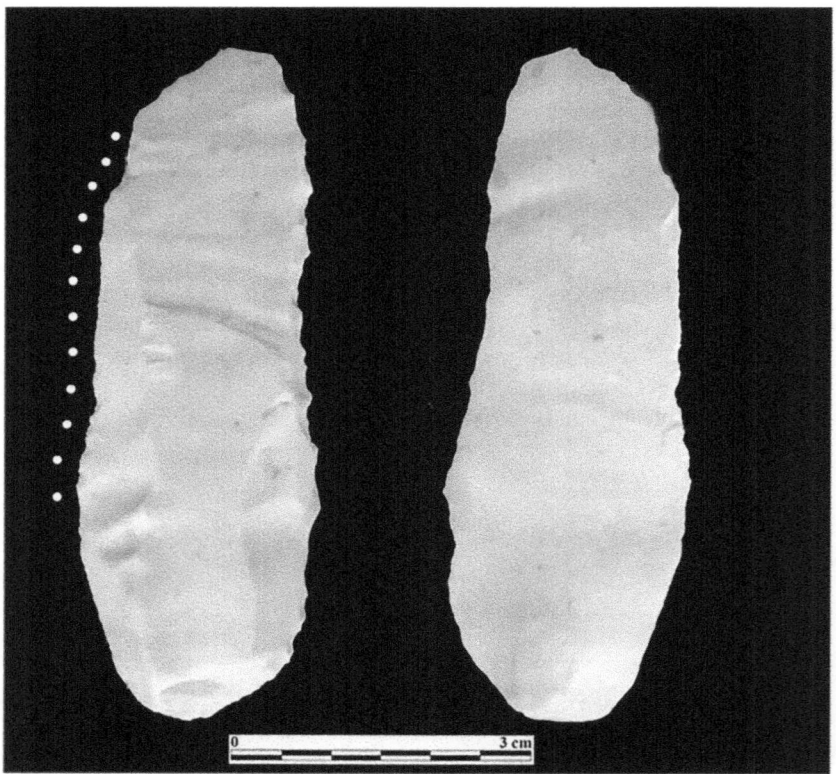

Fig. 7.4 Graphical documentation of traces (the section marks the microscopic image from Fig. 7.2; the dotted line shows the area with traces of use). (Graphic: Andreas Taller)

requisite for meaningful use-wear analyses (cf. Rots, 2010). The studies carried out are correspondingly extensive and wide-ranging (see above). *Blind tests* have become established as a means of verifying the results of experiments, as this allows the validity of the criteria and attributes that identify a particular method of working or a particular material that has been processed to be verified.

Post-depositional changes to lithic artefacts can complicate use-wear investigations. Such non-intentional, accidental traces may include edge rounding, *striae*, and polishing caused by transport in the sediment. Frost dynamics may cause chipping, and the chemical soil environment may cause patination. Finally, the handling of artefacts after their recovery can also generate microscopically visible traces (cf. Moss, 1983; Symens, 1988; Pawlik, 1995; Rots, 2010).

7.4   Strengths and Weaknesses

Ideally, both the working movement of a stone artifact and the worked material can be identified through use-wear analyses; and in part, statements can be made about the design of composite tools and hafting methods. The concrete function of an artefact can thus be ascertained, in contrast to purely technological analyses.

To achieve such results, however, it is first necessary to obtain a great deal of experience in microscopic analysis of stone artefacts, and there must be an in-depth study of comprehensive, well-documented and archived reference collections. Time-intensive experiments are essential to continually test the validity of the method and to be able to check hypotheses; in addition, the appropriate equipment must be available. The time-intensive nature of the analyses usually allows only for the analysis of numerically small samples of stone tools from entire lithic assemblages.

The long and intensive training period and the relatively time-consuming microscopic examination as such are weak points of the method, which is why it is usually limited to representative samples. The required equipment is relatively expensive to purchase.

What You Learned From This *essential*

- Gain of knowledge about the methodical range of the analysis of stone artefacts
- Selection of the appropriate analysis method for specific questions
- Combination of approaches of different methods
- Instructions and assistance for implementation
- Advice on the interpretation of the results

References

Anderson, P. C. (1980). A testimony of prehistoric tasks: Diagnostic residues on stone tool working edges. *World Archaeology, 12*(2), 181–194.

Anderson-Gerfaud, P. (1986). A few comments concerning residue analysis of stone plant-processing tools. In L. Owen & G. Unrath (Hrsg.), *Technical aspects of microwear studies on stone tools, early man news 9/10/11* (S. 69–81). Tübingen: Archaeologica Venatoria.

Andrefsky, W. (2005). *Lithics: Macroscopic approaches to analysis*. Cambridge University Press.

Audouze, F., & Karlin, C. (2017). La chaîne opératoire a 70 ans: qu'en ont fait les préhistoriens français. *Journal of Lithic Studies, 4*, 5–73.

Audouze, F., Bodu, P., Karlin, C., Julien, M., Pelegrin, J., & Perlès, C. (2017). Leroi-Gourhan and the chaîne opératoire: A response to Delage. *World Archaeology, 49*, 718–723.

Auffermann, B., Burkert, W., Hahn, J., Pasda, C., & Simon, U. (1990). Ein Merkmalsystem zur Auswertung von Steinartefaktinventaren. *Archäologisches Korrespondenzblatt, 20*, 259–268.

Baales, M., Birker, S., & Mucha, F. (2017). Hafting with beeswax in the final Palaeolithic: A barbed point from Bergkamen. *Antiquity, 91*, 1155–1170.

Balfet, H. (1975). Technologies. In R. Cresswell (Hrsg.), *Éléments d'ethnologie 2* (S. 44–79). A. Colin.

Bar-Yosef, O., & van Peer, P. (2009). The Chaîne Opératoire Approach in Middle Paleolithic Archaeology. *Current Anthropology, 50*(1), 103–131.

Bataille, G. (2006). The production and usage of stone artefacts in the context with faunal exploitation – The repeatedly visited primary butchering station of Level II/7E. In V. P. Chabai, J. Richter, & T. Uthmeier (Hrsg.), *Kabazi II: The 70000 years since the last interglacial. Palaeolithic sites of Crimea* (Bd. 2, S. 111–130). Shlyakh.

Bataille, G. (2010). Recurrent occupations of the late Middle Palaeolithic station Kabazi II, unit II, level 8 (Crimea, Ukraine) – Seasonal adaption, procurement and processing of resources. *Quartär, 57*, 43–77.

Bataille, G. 2012. Stones and Bones. The reconstruction of occupational palimpsests in the late Middle Palaeolithic of Crimea (Ukraine). In J. Cascalheira & C. Gonçalves (Hrsg.), *Actas das IV Jornadas de Jovens em Investigação Arqueológica – JIA 2011(2)*, 201–209. Promontoria Monográfica Bd. 16, Universidade do Algarve.

Bataille, G. (2016). Extracting the "Proto" from the Aurignacian. Dissociate and intercalated production sequences of blades and bladelets in the lower Aurignacian phase of Siuren 1, Units H & G (Crimea). *Mitteilungen der Gesellschaft für Urgeschichte, 25*, 49–86.

Bataille, G. (2017). Neanderthals of Crimea – Creative generalists of the late Middle Paleolithic. Contextualizing the leaf point industry Buran-Kaya III. *Level C. Quaternary International, 435*, 211–236.

Bataille G., Conard NJ (2018). Blade and bladelet production at Hohle Fels Cave, AH IV in the Swabian Jura and its importance for characterizing the technological variability of the Aurignacian in Central Europe. *PLoS ONE, 13*(4), e0194097.

Beyries, S. (1993). Exprérimentation archéologique et savoir-faire traditionnel: l´exemple de la découpe d´un cervidé. *Techniques et cultures, 22*, 53–79.

Beyries, S. (1999). Ethnoarchaeology: A Method of Experimentation. *Urgeschichtliche Materialhefte, 14*, 117–130.

Boëda, E. (1986). *Approche Technologique du Concept Levallois et Evaluation de son Champ d'Application*. Université Paris X-Nanterre.

Boëda, E. (1988). Le concept laminaire: rupture et filiation avec le concept Levallois. In M. Otte (Hrsg.), *L'homme de Néandertal* (S. 41–59). Liège: ERAUL 8.

Boëda, E. (1994). *Le concept Levallois: variabilité des méthodes*. C.N.R.S. Monographie du CRA.

Boëda, E. (1995). Steinartefakt-Produktionssequenzen im Micoquien der Kulna-Höhle. *Quartär, 45*(46), 75–98.

Boëda, E. (1997). *Technogenèse de systèmes de production lithique au Paléolithique inférieur et moyen en Europe occidentale et au Proche-Orient*. Université Paris X-Nanterre (Habilitation).

Boëda, E. (2001). Determination des unites techno-fonctionnelles de pieces bifaciales provenant de la couche acheuléenne C'3 base du site de Barbas I. In D. Cliquet (Hrsg.), *Les industries à outils bifaciaux du Paléolithique moyen d'Europe occidentale. Actes de la Table Ronde internationale de Caen, 14–15 octobre 1995* (S. 51–75). ERAUL 98.

Boëda, E. (2013). *Technologique & technologie: Une paléo-histoire des objets lithiques tranchants*. Archéo-éditions.

Boëda, E., Geneste, J. M., & Meignen, L. (1990). Identification de chaînes opératoires lithiques du Paléolithique ancien et moyen. *Paléo, 2*, 43–80.

Böhner, U. (2008). *Die Schicht E3 der Sesselfelsgrotte und die Funde aus dem Abri I am Schulerloch. Späte Micoquien-Inventare und ihre Stellung zum Moustérien* (Bd. IV). Verlag Franz Steiner.

Bordes, F. (1950). Principes d'une méthode d'étude des techniques et de la typologie du Paléolithique. ancien et moyen. *L'Anthropologie, 54*, 19–34.

Bordes, F. (1961). *Typologie du Paléolithique ancien et moyen. Mémoire n°1*. Publications de l'Institut de Préhistoire de l'Université de Bordeaux.

Bosinski, G., Brunnacker, K., Schütrumpf, R., & Rottländer, R. (1966). Der paläolithische Fundplatz Rheindahlen, Ziegelei Dreesen-Westwand. *Bonner Jahrbücher, 166*, 318–360.

Bourguignon, L. (1992). Analyse du processus opératoire des coups de tranchet lateraux dans l'industrie moustérienne de l'Abri du Musée (Les Eyzies-de-Tayac, Dordogne). *Paléo, 4*, 69–89.

Bradtmöller, M., Sarmiento, A., Perales, U., & Cruz Zuluaga, M. (2016). Investigation of Upper Palaeolithic adhesive residues from Cueva Morín, Northern Spain. *Journal of Archaeological Science Reports, 7*, 1–13.

Briuer, F. (1976). New clues to stone tool function: Plant and animal residues. *American Antiquity, 41*(4), 478–484.

Burkert, W. (1999). *Lithische Rohmaterialversorgung im Jungpaläolithikum des südöstlichen Baden-Württemberg*. Dissertation, Tübingen.

Caspar, J.-P., & de Bie, M. (1996). Preparing for the Hunt in the Late Paleolithic Camp at Rekem. *Belgium. Journal of Field Archaeology, 23*(4), 437–460.

Çep, B., Burkert, W., & Floss, H. (2011). Zur mittelpaläolithischen Rohmaterialversorgung im Bockstein (Schwäbische Alb). *Mitteilungen der Gesellschaft für Urgeschichte, 20*, 33–46.

Chabai, V. P., Richter, J., & Uthmeier, T. (2005). *Kabazi II: Last interglacial occupation, environment & subsistence. Palaeolithic sites of Crimea* (Bd. 1). Shlyakh.

Chabai, V. P., Richter, J., & Uthmeier, T. (2006). *Kabazi II: The 70000 years since the last interglacial. Palaeolithic sites of Crimea* (Bd. 2). Shlyakh.

Chomsky, N. (1965). *Aspects of the theory of syntax*. MIT Press.

Cnurs, D., Perrault, K. A., Stefanuto, P.-H., Dubois, L. M., Focant, J.-F., & Rots, V. (2018). Fingerprinting glues using HS-SPME GC×GC-HRTOFMS: A new powerful method allows tracking glues back in time. *Archaeometry, 60*(6), 1361–1376.

Conard, N. J., Prindiville, T. J., & Adler, D. S. (1998). Refitting bones and stones as a means of reconstructing middle paleolithic subsistence in the Rhineland. In J. P. Brugal, L. Meignen, & M. Patou-Mathis (Hrsg.), *XVIlle Rencontres Internationales d'Archéologie et d'Histoire d'Antibes, Économie Préhistorique: Les comportements de subsistence au Paléolithique* (S. 273–290). Éditions APDCA.

Cotterell, B., & Kamminga, J. (1987). The formation of flakes. *American Antiquity, 52*, 675–708.

Cotterell, B., Kamminga, J., & Dickson, F. P. (1985). The essential mechanics of conchoidal flaking. *International Journal of Fracture, 29*(4), 205–221.

Cresswell, R. (1983). Transferts de techniques et chaînes opératoires. *Techniques et Culture, 2*, 143–163.

Cziesla, E. (1986). Über das Zusammenpassen geschlagener Steinartefakte. *Archäologisches Korrespondenzblatt, 16*, 251–265.

Dauvois, M. (1976). *Précis de dessin dynamique et structural des industries lithiques préhistoriques*. Pierre Fanlac.

Dibble, H., & Bernard, M. C. (1980). A comparative study of basic edge angle measurement techniques. *American Antiquity, 45*(4), 857–865.

Dibble, H. L., & Rezek, Z. (2009). Introducing a new experimental design for controlled studies of flake formation: Results for exterior platform angle, platform depth, angle of blow, velocity, and force. *Journal of Archaeological Science, 36*(9), 1945–1954.

Dibble, H. L., & Whittaker, J. C. (1981). New experimental evidence on the relation between percussion flaking and flake variation. *Journal of Archaeological Science, 8*, 283–296.

Donnart, K. (2010). L'analyse des unités techno-fonctionnelles appliquée à l'étude du macro-outillage néolithique. *L'Anthropologie, 114*(2), 179–198.

Drafehn, A., Bradtmöller, M., & Mischka, D. (2008). SDS-Systematische und digitale Erfassung von Steinartefakten (Arbeitsstand SDS 8.05). *Journal of Neolithic Archaeology, 10*, 1–58.

Fischer, A., Hansen, P. V., & Rasmussen, P. (1984). Macro and micro wear traces on lithic projectile points. Experimental results and prehistoric examples. *Journal of Danish Archaeology, 3*, 19–46.

Fish, P. R. (1981). Beyond tools: Middle Paleolithic debitage analysis and cultural inference. *Journal of Anthropological Research, 37*(4), 374–386.

Floss, H. (1994). *Rohmaterialversorgung im Paläolithikum des Mittelrheingebietes.* Monographien des RGZM 21. Bonn: Rudolf Habelt Verlag.

Floss, H. (Ed.). (2012). *Steinartefakte. Vom Altpaläolithikum bis in die Neuzeit. Tübingen Publications in Prehistory.* Kerns Verlag.

Frick, J. A., Herkert, K., Hoyer, C., & Floss, H. (2017). The performance of tranchet blows at the late Middle Palaeolithic site of Grotte de la Verpillère I (Saône-et-Loire, France). *PlosOne, 12*(11), e0188990.

Geneste, J.-M. (1985). *Analyse lithique d'industries moustériennes du Périgord: approche technologique du comportement des groupes humains au Paléolithique moyen.* Université de Bordeaux I (Unpublizierte Doktorarbeit).

Geneste, J.-M. (1991). Systèmes techniques de production lithique: variations techno-économiques dans les processus de réalisation des outillages paléolithiques. *Techniques et Culture. 17–18*, 1–35.

Hahn, J. (1988). *Die Geißenklösterle-Höhle im Achtal bei Blaubeuren: Fundhorizontbildung und Besiedlung im Mittelpaläolithikum und im Aurignacien* (Bd. 26). Forschungen und Berichte zur Vor- und Frühgeschichte in Baden-Württemberg. K. Theiss.

Hahn, J. (1991). *Erkennen und Bestimmen von Stein- und Knochenartefakten: Einführung in die Artefaktmorphologie* (Bd. 10). Archaeologica Venatoria.

Hahn, J. (1992). *Zeichnen von Stein- und Knochenartefakten* (Bd. 13). Archaeologica Venatoria.

Hassan, F. A. (1988). Prolegomena to a grammatical theory of lithic artifacts. *World Archaeology, 19*(3), 281–296.

Haudricourt, A.-G. (1964). La technologie, science humaine. *La Pensée, 115*, 28–35.

Haudricourt, A.-G. (1987). *La technologie, science humaine, Recherche d'histoire et d'ethnologie des techniques.* Maison des Sciences de l'Homme.

Hardy, B. L. & Garufi, G. T. (1998). Identification of Woodworking on Stone Tools through Residue and Use-Wear Analyses: Experimental Results. *Journal of Archaeological Science, 25*(2), 177–184.

Hayden, B. (Ed.). (1979). *Lithic use-wear analysis.* Academic Press.

Herzog, I. (2010). Stratify website. http://www.stratify.org/.

Holdaway, S., & Stern, N. (2004). *A record in stone: The study of Australia's flaked stone artifacts.* Aboriginal Studies Press.

Inizan, M.-L., Roche, H., & Tixier, J. (1992). *Technology of Knapped Stone. Préhistoire de la Pierre Taillée, t. 3.* CREP.

Inizan, M.-L., Reduron-Ballinger, M., Roche, H., & Tixier, J. (1995). *Préhistoire de la Pierre Taillée – t. 4: Technologie de la pierre taillée.* CREP.

Inizan, M.-L., Reduron-Ballinger, M., Roche, H., & Tixier, J. (1999). *Technology and terminology of knapped stone.* CREP.

Jöris, O. (2001). *Der spätmittelpaläolithische Fundplatz Buhlen (Grabungen 1966–1969). Stratigraphie, Steinartefakte und Fauna des Oberen Fundplatzes* (Bd. 73). Universitätsforschungen zur Prähistorischen Archäologie. Habelt.

Keeley, L. H. (1974). Technique and methodology in microwear studies – Critical review. *World Archaeology, 5*(3), 323–336.

Keeley, L. H. (1980). *Experimental determination of stone tool uses: A microwear analysis.* University of Chicago Press.

Keeley, L. H. (1982). Hafting and retooling: Effects on the archaeological record. *American Antiquity, 47*(4), 798–809.

Keeley, L. H., & Newcomer, M. H. (1977). Microwear analysis of experimental flint tools: A test case. *Journal of Archaeological Science, 4,* 29–62.

Kerkhof, F., & Müller-Beck, H. (1969). Zur bruchmechanischen Deutung der Schlagmarken an Steingeräten. *Glastechnische Berichte, 42,* 439–448.

Kind, C.-J. (2003). *Das Mesolithikum in der Talaue des Neckars. Die Fundstellen von Rottenburg Siebenlinden 1 und 3* (Bd. 88). Forschungen und Berichte zur Vor- und Frühgeschichte in Baden-Württemberg. Konrad Theiss Verlag.

Kretschmer, I. (2006). Kabazi II, Level II/7AB: Hunting and raw material procurement for stone artefact production. In V. P. Chabai, J. Richter, & T. Uthmeier (Hrsg.), *Kabazi II: The 70000 years since the last interglacial. Palaeolithic sites of Crimea* (Bd. 2, S. 73–83). Simferopol: Shlyakh.

Kurbjuhn, M. (2005). Operational sequences of bifacial production in Kabazi II, units V and VI. In V. P. Chabai, J. Richter, & T. Uthmeier (Hrsg.), *Kabazi II: Last interglacial occupation, environment and subsistence. Palaeolithic sites of Crimea* (Bd. 1, S. 257–274). Simferopol: Shlyakh.

Lepot, M. (1993). *Approche techno-fonctionelle de l'outillage moustérien. Essai de classification des parties actives en terme d'éfficacité technique. Application à la couche M2e sagittale du grand abri de la Ferrassie (fouille Delporte).* Université Paris X-Nanterre (Mémoire de Maîtrise).

Leroi-Gourhan, A. (1943). *Evolution et technique I – L'Homme et la Matière.* Albin Michel.

Leroi-Gourhan, A. (1964). *Le geste et la parole I - Technique et language.* Albin Michel.

Löhr, H. (1979). *Der Magdalénien-Fundplatz Alsdorf, Kreis Aachen-Land. Ein Beitrag zur Kenntnis der funktionalen Variabilität jungpaläolithischer Stationen.* Dissertation, Tübingen.

Lombard, M., & Wadley, L. (2007). The morphological identification of micro-residues on stone tools using light microscopy: Progress and difficulties based on blind tests. *Journal of Archaeological Science, 34,* 155–165.

Loy, T. H. (1993). The artifact as site: An example of the biomolecular analysis of organic residues on prehistoric tools. *World Archaeology, 25*(1), 44–63.

Machado, J., Molina, F. J., Hernández, C. M., Tarriño, A., & Galván, B. (2016). Using lithic assemblage formation to approach middle palaeolithic settlement dynamics: El Salt stratigraphic unit X (Alicante, Spain). *Archaeological and Anthropological Sciences, 9,* 715–1743.

Magne, M. P. R. (1985). *Lithics and livelihood: Stone tool technologies of central and southern interior British Columbia. Mercury Series No. 133.* National Museum of Man.

Marreiros, J., Gibaja, J. F., & Bao, N. F. B. (Eds.). (2014). *Use-Wear and Residue Analysis in Archaeology*. Springer.

Mauss, M. (1947). *Manuel d'ethnographie*. Payot.

Moss, E. H. (1983). *The functional analysis of flint implements. Pincevent and Pont d'Ambon: Two case studies from the french final palaeolithic* (Bd. 177). BAR International Series.

Moss, E. H., & Newcomer, M. H. (1982). Reconstruction of tool use at Pincevent: Microwear and experiments. *Studia Praehistorica Belgica, 2*, 289–312.

Nance, J. D. (1971). Functional interpretation from microscopic analysis. *American Antiquity, 36*, 361–366.

Odell, G. H. (2004). *Lithic analysis. Manuals in archaeological method, theory, and technique*. Kluwer Academic.

Odell, G. H., & Odell-Vereecken, F. (1980). Verifying the reliability of lithic use-wear assessments by 'Blind Tests': The low-power approach. *Journal of Field Archaeology, 7*(1), 87–120.

Ollé, A., & Verges, J. M. (2014). The use of sequential experiments and SEM in documenting stone tool microwear. *Journal of Archaeological Science, 48*, 60–72.

Pastoors, A. (2000a). Standardization and Individuality in the production process of bifacial tools – Leaf-shaped scrapers from the Middle Paleolithic open air site Saré Kaya I (Crimea). A contribution to understanding the method of Working Stage Analysis. In J. Orschiedt & G. C. Weniger (Hrsg.), *Neanderthals and modern humans – Discussing the transition. Central and eastern Europe from 50.000 – 30.000 B.P.* 243–255 (Bd. 2). Wissenschaftliche Schriften des Neanderthal Museums. Neanderthal Museum.

Pastoors, A. (2000b). Normierung und Individualität im Herstellungsprozess bifazialer Werkzeuge – Blattförmige Schaber von der mittelpaläolithischen Freilandstation Saré Kaya I (Krim): Ein Beitrag zum Verständnis der Arbeitsschrittanalyse: Grundlagen. *Anwendung und Auswertung. Archäologisches Korrespondenzblatt, 30*(2), 153–164.

Pastoors, A. (2001). *Die mittelpaläolithische Freilandstation von Salzgitter-Lebenstedt. Genese der Fundstelle und Systematik der Steinbearbeitung* (Bd. 3). Salzgitter Forschungen. Ruth Printmedien.

Pastoors, A., & Schäfer, J. (1999). Analyse des états techniques de transformation, d'utilisation et états post dépositionnels. Illustrée par un outil bifacial de Salzgitter-Lebenstedt (FRG). *Préhistoire Européenne, 14*, 33–47.

Pastoors, A., Tafelmaier, Y., & Weniger, G.-C. (2015). Quantification of late pleistocene core configurations: Application of the working stage analysis as estimation method for technological behavioural efficiency. *Quartär, 62*, 63–84.

Pawlik, A. (1995). *Die mikroskopische Analyse von Steingeräten. Experimente – Auswertungsmethoden – Artefaktanalysen* (Bd. 10). Archaeologica Venatoria.

Pelcin, A. W. (1997). The formation of flakes: The role of platform thickness and exterior platform angle in the production of flake initiations and terminations. *Journal of Archaeological Science, 24*(12), 1107–1113.

Pelegrin, J. (1990). Prehistoric lithic technology: Some aspects of research. *Archaeological Review from Cambridge, 9*, 116–125.

Pelegrin, J. (1995). *Technologie lithique: Le Châtelperronien de Roc-de-Combe (Lot) et de la Côte (Dordogne)*. Cahiers du Quaternaire. Édition du CNRS.

Pelegrin, J. (2000). Les techniques de débitage laminaire au Tardiglaciaire: critères de diagnose et quelques réflexions. In B. Valentin, P. Bodu, & M. Christensen (Hrsg.), *L'Europe*

Centrale et Septentrionale au Tardiglaciaire. Confrontation des modèles régionaux de peuplement (Bd. 7). Mémoires de Musée de Préhistoire d'Ile de France (S. 73–86). APRAIF.

Perlès, C. (1989). *Les industries lithiques taillées de Franchthi (Argolide, Grèce), tome 1: Présentation générale et industries Paléolithique.* Indiana University Press.

Perrault, K., Stefanuto, P.-H., Dubois, L., Cnuts, D., Rots, V., & Focant, J.-F. (2016). A new approach for the characterization of organic residues from stone tools using GC×GC-TOFMS. *Separations, 3*(2), 1–13.

Pétillon, J.-M., Bignon, O., Bodu, P., Cattelain, P., Debout, G., Langlais, M., Laroulandie, V., Plisson, H., & Valentin, B. (2011). Hard core and cutting edge: Experimental manufacture and use of Magdalenian composite projectile tips. *Journal of Archaeological Science, 38*, 1266–1283.

Pigeot, N. (1991). Réflexions sur l'histoire technique de l'homme: de l'évolution cognitive à l'évolution culturelle. *Paléo, 3*, 167–200.

Plisson, H. (1985). *Étude fonctionelle d'outillages lithiques préhistoriques par l'analyse des micro-usures: recherche méthodologique et archéologique.* Université Paris 1-Panthéon Sorbonne (Unpublizierte Doktorarbeit).

Porraz, G. (2005). *En marge du milieu alpin – Dynamiques de formation des ensembles lithiques et modes d'occupation des territoires au Paléolithique moyen.* Aix-Marseille Université I (Unpublizierte Doktorarbeit).

Porraz, G., Igreja, M., Schmidt, P., & Parkington, J. E. (2016). A shape to the microlithic Robberg from Elands Bay Cave (South Africa). *Southern African Humanities, 29*, 203–247.

Rabardel, P. (1995). *Les hommes et les technologies; approche cognitive des instruments contemporains.* Armand Colin Éditeurs.

Rezek, Z., Lin, S. C., Iovita, R. P., & Dibble, H. L. (2011). The relative effects of core surface morphology on flake shape and other attributes. *Journal of Archaeological Science, 38*, 1346–1359.

Richter, J. (1997). *Sesselfelsgrotte III: Der G-Schichten-Komplex der Sesselfelsgrotte – Zum Verständnis des Micoquien* (Bd. 7). Quartär-Bibliothek. Saarbrücker Druckerei.

Richter, J. (2001). Une analyse standardisée des chaînes opératoires sur les pièces foliacées du Paléolithique moyen tardif. In L. Bourguignon, I. Ortega, & M.-C. Frère-Sautot (Hrsg.), Préhistoire et approche expérimentale. *Préhistoires 5* (S. 77–87). Éditions Mergoil.

Richter, J. (2018). *Altsteinzeit. Der Weg der frühen Menschen von Afrika bis in die Mitte Europas.* Verlag W. Kohlhammer.

Roebroeks, W. (1988). *From find scatters to early hominid behavior. A study of Middle Palaeolithic riverside settlements at Maastricht-Belvédère (The Netherlands).* Analecta Praehistorica Leidensia 21.

Romagnoli, F., & Vaquero, M. (2016). Quantitative stone tools intra-site point and orientation patterns of a Middle Palaeolithic living floor: A GIS multi-scalar spatial and temporal approach. *Quartär, 63*, 47–60.

Romagnoli, F., Bargalló, A., Chacón, M. G., Gómez de Soler, B., & Vaquero, M. (2016). Testing a hypothesis about the importance of the quality of raw material on technological changes at Abric Romaní (Capellades, Spain): Some considerations using a high-resolution techno-economic perspective. *Journal of Lithic Studies, 3*(2), 1–25.

Rots, V. (2010). *Prehension and Hafting traces on flint tools. A methodology.* Leuven University Press.

Rots, V., & Plisson, H. (2014). Projectiles and the abuse of the use-wear method in a search for impact. *Journal of Archaeological Science, 48,* 154–165.

Rots, V., Hardy, B. L., Serangeli, J., & Conard, N. J. (2015). Residue and microwear analyses of the stone artifacts from Schöningen. *Journal of Archaeological Science, 89,* 298–308.

Schlanger, N. (1991). Le fait technique total. La raison pratique et les raisons de la pratique dans l'oeuvre de Marcel Mauss. *Association Terrain, 16,* 114–130.

Semenov, S. A. (1964). *Prehistoric technology – An experimental study of the oldest tools and artefacts from traces of manufacture and wear.* Adams & Dart.

Shea, J. J. (2013). *Stone tools in the Paleolithic and Neolithic near east: A guide.* Cambridge University Press.

Shott, M. J. (1994). Size and form in the analysis of flake debris: Review and recent approaches. *Journal of Archaeologial Method and Theory, 1,* 69–110.

Soressi, M. (2002). Le Moustérien de tradition acheuléenne du sud-ouest de la France. Discussion sur la signification du faciès à partir de l'étude comparée de quatre sites: Pech-de-l'Azé I, Le Moustier, La Rochette et la Grotte XVI. PhD thesis, Université Bordeaux I.

Soressi, M., & Geneste, J.-M. (2011). Special issue: Reduction sequence, chaîne opératoire, and other methods: The epistemologies of different approaches to lithic analysis. The history and efficacy of the chaîne opératoire approach to lithic analysis: Studying techniques to reveal past societies in an evolutionary perspective. *PaleoAnthropology, 2011,* 334–350.

Soriano, S. (2001). Statut fonctionnel de l'outillage bifacial dans les industries du Paléolithique moyen: proposition méthodologique. In D. Cliquet (Hrsg.), *Les industries à outils bifaciaux du Paléolithique moyen d'Europe occidentale. Actes de la Table Ronde internationale de Caen* (S. 77–83). ERAUL 98.

Speth, J. (1972). The mechanical basis of percussion flaking. *American Antiquity, 37,* 34–60.

Spurrell, F. (1884). On some palaeolithic knapping tools and modes of using them. *Journal of the Royal Anthropological Institute of Great Britain and Ireland, 13,* 109–118.

Sullivan, A. P., & Rozen, K. C. (1985). Debitage analysis and archaeological interpretation. *American Antiquity, 50,* 755–779.

Symens, N. 1988. Mikroskopische Analyse der Oberfläche von Steinartefakten. In J. Hahn (Hrsg.), *Die Geißenklösterle-Höhle im Achtal bei Blaubeuren 1,* Forschungen und Berichte zur Vor- und Frühgeschichte in Baden-Württemberg 26 (S. 59–201). Theiss eVerlag.

Tafelmaier, Y. (2010). *Das steinzeitliche Fundmaterial der Volkringhauser Höhle im Hönnetal/Westfalen.* Universität zu Köln (Unpubulizierte Magisterarbeit).

Tafelmaier, Y. (2011). Revisiting the Middle Palaeolithic site Volkringhauser Höhle (North Rhine-Westphalia, Germany). *Quartär, 58,* 153–182.

Thieme, H. (1983). *Der paläolithische Fundplatz Rheindahlen.* Inaugural-Dissertation, Köln.

Tixier, J. (1967). Procédés d'analyse et questions de terminologie dans l'étude des ensembles industriels du Paléolithique récent et de l'Epipaléolithique en Afrique du Nord-Ouest. In W. W. Bishop & J. D. Clark (Hrsg.), *Background to evolution in Africa* (S. 771–820). University of Chicago Press.

Tixier, J. (1980). *Préhistoire et technologie lithique.* Édition du CNRS.

Tixier, J. (2012). *A method for the study of stone tools = méthode pour l'étude des outill-ages lithiques: guidelines based on the work of J. Tixier = notice sur les travaux sci-entifiques de J. Tixier.* Musée national d'histoire et d'art/Centre national de recherche archéologique du Luxembourg.

Tixier, J., Inizan, M.-L., & Roche, H. (1980). *Préhistoire de la Pierre Taillée 1: Terminologie et Technologie.* Cercle de Recherches et d'Etudes Préhistoriques.

Tostevin, G. B. (2003). Attribute analysis of the lithic technologies of Stránská Skála II–III in their regional and inter-regional context. In J. Svoboda & O. Bar-Yosef (Hrsg.), *Stránská Skála: Origins of the upper palaeolithic in the brno basin* (S. 77–118). Peabody Museum Publications.

Tostevin, G. B. (2012). *Seeing lithics: A middle-range theory for testing cultural transmis-sion in the Pleistocene.* Oxbow Books.

Unrath, G., Owen, L. R., van Gijn, A., Moss, E. H., Plisson, H., & Vaughan, P. (1986). An evaluation of use-wear studies: A multi-analyst approach. In L. Owen & G. Unrath (Hrsg.), *Technical aspects of microwear studies on stone tools, early man news 9/10/11* (S. 117–176). Archaeologica Venatoria.

Uthmeier, T. (2004a). Transformation analysis and the reconstruction of on-site and off-site activities: Methodological remarks. In V. P. Chabai, K. Monigal, & A. E. Marks (Hrsg.), *The middle Paleolithic and early upper Paleolithic of eastern Crimea. The Paleolithic of Crimea III* (Bd. 104, S. 175–191). ERAUL.

Uthmeier, T. (2004b). Planning depth and saiga hunting: On-site and off-site activities of late Neanderthals. In V. P. Chabai, K. Monigal, & A. E. Marks (Hrsg.), *The middle Paleolithic and early upper Paleolithic of eastern Crimea. The Paleolithic of Crimea III* (Bd. 104, S. 193–231). ERAUL.

Uthmeier, T. (2004c). *Micoquien, Aurignacien und Gravettien in Bayern. Eine regionale Studie zum Übergang vom Mittel- zum Jungpaläolithikum.* Bd. 18: Archäologische Berichte. Bonn: Rudolf Habelt.

Vaughan, P. (1985). *Use-wear analysis of flaked stone tools.* University of Arizona Press.

Weißmüller, W. (1995). *Die Silexartefakte der Unteren Schichten der Sesselfelsgrotte. Ein Beitrag zum Problem des Moustérien* (Bd. 6). Quartär-Bibliothek. Saarbrücken: Saar-brücker Druckerei.

Whittaker, J. C. (1994). *Flintknapping: Making and understanding stone tools.* University of Texas Press.

Wilmsen, E. N. (1968). Functional analysis of flaked stone artifacts. *American Antiquity, 33*(2), 156–161.

Yates, A. B., Smith, A. M., Bertuch, F., Gehlen, B., Gramsch, B., Heinen, M., Joannes-Boyau, R., Scheffers, A., Parr, J., & Pawlik, A. (2015). Radiocarbon-dating adhesive and wooden residues from stone tools by Accelerator Mass Spectrometry (AMS): Challenges and insights encountered in a case study. *Journal of Archaeological Science, 61*, 45–58.

Zimmermann, A. (1988). Steinmaterial. In U. Boelicke, D. von Brandt, J. Lüning, P. Stehli, & A. Zimmermann (Hrsg.), *Der Bandkeramische Siedlungsplatz Langweiler 8, Gem. Aldenhoven, Kr. Düren* (Bd. 28, S. 569–787). Beiträge zur neolithischen Besiedlung der Aldenhovener Platte III. Rheinische Ausgrabungen.

GPSR Compliance
The European Union's (EU) General Product Safety Regulation (GPSR) is a set
of rules that requires consumer products to be safe and our obligations to
ensure this.

If you have any concerns about our products, you can contact us on

ProductSafety@springernature.com

In case Publisher is established outside the EU, the EU authorized
representative is:

Springer Nature Customer Service Center GmbH
Europaplatz 3
69115 Heidelberg, Germany